Thomas C. MacGinley

An Introduction to the Study of General Biology

Designed for the Use of Schools and Science Classes

Thomas C. MacGinley

An Introduction to the Study of General Biology
Designed for the Use of Schools and Science Classes

ISBN/EAN: 9783337216900

Printed in Europe, USA, Canada, Australia, Japan

Cover: Foto ©Paul-Georg Meister /pixelio.de

More available books at **www.hansebooks.com**

Putnam's Elementary Science Series.

AN INTRODUCTION

TO THE STUDY OF

GENERAL BIOLOGY.

DESIGNED FOR THE USE OF

SCHOOLS AND SCIENCE CLASSES.

BY

THOMAS C. MacGINLEY,

PRINCIPAL, CROAGH NATIONAL SCHOOL, COUNTY DONEGAL.

WITH 124 ILLUSTRATIONS.

NEW YORK:
G. P. PUTNAM'S SONS,
FOURTH AVENUE AND TWENTY-THIRD STREET

PREFACE.

THE very best preparation for the study of the laws of LIFE, under its various manifestations, is first to become thoroughly conversant with the characters of a few well-marked typical beings, with which all others may be compared, and around which, as centres, they may be conveniently grouped. As every true classification of plants and animals must be founded on their structural affinities, care has been taken in the present work to place before the student a clearly detailed statement of the form, functions, and minute structure of every part of the organism brought under his notice. With this object in view, numerous illustrations have been introduced, which, it is hoped, will render the facts of anatomy and physiology more intelligible, and more easily remembered. The student is, however, earnestly recommended to verify these by actual observation. The illustrations, for the most part, are either drawings from nature, or *diagrams* of the Author's own designing. A few, however, are taken from the works of recent writers on the subjects to which they severally refer. The use of technical terms has been by no means avoided, inasmuch as it is believed to be absolutely impossible to understand thoroughly any department of special study without a good knowledge of the technical language employed by the more eminent among the authorities who treat of it. In the present work, these terms are explained as they occur, and in the Glossary their derivations and literal meanings are also given along with their technical applications.

In the preparation of this work, designed, as it is, to meet the wants of junior science students in connection with the Science and Art Department, the Syllabus of the Elementary Stage in General Biology has been followed. It is not intended, however, to supersede the use of the scalpel or microscope, or oral instruction from a competent teacher. On the contrary, it is hoped that, with its aid, and with the other helps recommended, the student will examine the great field of nature for himself, and thus acquire that kind of knowledge which it is most desirable and delightful to possess—namely, the knowledge derived from a patient questioning and examination of those objects in nature to which one's studies are specially devoted.

Of the works consulted, the author must here express his indebtedness to those of Balfour, Silver, Haughton, Carpenter, Nicholson, and Huxley. He must also acknowledge his obligations to Messrs. P. Doyle, and G. H. Begley, Portaferry, and to Messrs. J. C. Ward and P. O'Byrne, Killybegs, for the aid kindly rendered while the work was passing through the press.

THOMAS C. MacGINLEY.

CROAGH, DUNKINEELY, DONEGAL,
March, 1874.

CONTENTS

CHAPTER I.
THE TORULA.

Introduction — Plasma of Torula — Pasteur's Fluid— Morphology — Protoplasm — Protein — Reproduction — Collateral Results of Torula's Existence — Fermentation — Production of Carbonic Acid and Alcohol — Theories, - - - - - - 9

CHAPTER II.
THE PROTOCOCCUS.

Plasma — Morphology — Mobile Forms — Physiology — Growth and Development — Protococcus and Torula Compared — Action of Chlorophyll under Sunlight — Cilia, - - - - - - - 16

CHAPTER III.
THE AMŒBA AND THE COLOURLESS CORPUSCLES OF THE BLOOD.

Plasma of Amœba — Morphology — Physiology — Conditions of Existence — Colourless Corpuscles of Human Blood: their Origin and Functions — Morphology, Chemical Composition, and Physiology of Torula, Protococcus, and Amœba (comparative view), - - - 20

CHAPTER IV.
THE BACTERIUM.

Plasma — Morphology — Brownian Movement — Reproduction — Putrefaction — Experiments — Spontaneous Generation, - - - - - - 26

CHAPTER V.

THE PENICILLIUM.

Plasma—Morphology—Hyphæ—Mycelium—Conditions of Existence — Reproduction — Peronospore — Potato Blight—Conjugation—Achlya—Fungi, their Distinctive Characteristics—Mushroom—Histology,   29

CHAPTER VI.

THE CHARA.

Plasma—Morphology—Segmentation—Homology of Parts—Nodes and Internodes—Primordial Utricle—Chlorophyll—Cyclosis—Physiology—Mode of Growth—Reproduction—Summary,   37

CHAPTER VII.

THE FERN.

Morphology—Rhizome or Root-Stock—Axis—Tissues—The Frond: its Mode of Growth—Absorption and Decomposition of Carbonic Acid—Stomata—Fixation of Carbon—Evolution of Oxygen—Nutrition—Exhalation and Inhalation of Moisture — Reproduction, Asexual and Sexual—Spores—Prothallium—Embryo—Alternation of Generations,   45

CHAPTER VIII.

THE BEAN (MORPHOLOGY).

Axis and Appendages—Flower—Stem—Stomata—Intercellular Spaces—Tissues,   54

CHAPTER IX.

THE BEAN (PHYSIOLOGY).

Cell Growth—Growth in Stem, Root, and Leaves—Veins—Branches—Flower — Stamens—Carpel — Fertilization of Ovule—Seed—Gamogenesis—Asexual Propagation (Agamogenesis)—Flowering Plant Compared with Fern,   60

CHAPTER X.

NUTRITION IN PLANTS.

Cells and Intercellular Spaces—Sources of Nutriment—Organic Matter in Soil not Essential—Functions of Chlorophyll—Ascent of Crude Sap—Endosmose—Elaboration and Distribution of Sap, - - - 71

CHAPTER XI.

EXOGENS AND ENDOGENS—MODIFICATIONS OF LEAF.

Exogens—Pith—Arrangement of Fibro-vascular Bundles—Medullary Rays—Cambium—Bark—Alburnum or Sap Wood—Duramen or Hard Wood—Endogenous Stem—Mode of Growth—Disposition of Fibro-vascular Bundles—Roots—Leaves in Exogens—Modifications of Leaf—Stipules—Bracts—Bud-scales—Spines—Tendrils—Phyllodes—Pitchers—Venation, Reticulated and Parallel—Differences between Exogens and Endogens, - - - - - - 75

CHAPTER XII.

THE FRESH-WATER POLYPE (HYDRA), AND THE SEA-ANEMONE (ACTINIA).

Cœlenterata — Their Distinctive Characters. *Hydra*—Morphology—Thread Cells—Muscular Fibres—Physiology—Alimentation—Irritability—Locomotion—Reproduction, Asexual and Sexual. *Actinia*—Morphology—Irritability—Physiology—Locomotion—Reproduction, Asexual and Sexual—Coralligena—Coral Reefs, - - - - - - - 82

CHAPTER XIII.

THE FRESH-WATER MUSSEL (ANODON).

Morphology—Pallium—Foot—Organ of Bojanus—Gills—Valves—Pallial and Adductor Impressions—Physiology—Alimentary System—Liver—Blood Circulation—Heart—Pericardium—Functions of Organ of Bojanus—Respiration — Locomotion — Muscular System—Nervous System—Ganglia—Reproduction, - - 98

CHAPTER XIV.

THE LOBSTER (AN ARTHROPOD).

Morphology—Somites—Appendages—Basipodite, Endopodite, Exopodite, Epipodite—Abdomen—Cephalothorax—Carapace—Branchiostegite—Gills—Ambulatory Limbs—Foot-jaws—Enumeration of Somites and Appendages—Structure of Appendages—Heart—Arteries. Physiology—Alimentation—Muscular Fibre of Intestines—Liver—Circulation—Heart and Pericardium—Aortas—Gills—Respiration—Green-Glands—Muscular System—Muscular Tissue—Peculiar Action of Muscles in Flexing Joints—Nervous System—Œsophageal Collar—Ganglia—Sensory Organs (Eyes, Ears)—Reproduction—Zoea—Cray-Fish, ... 113

CHAPTER XV.

THE FROG (AN AMPHIBIAN VERTEBRATE).

Morphology—External Form—Internal Structure—Histology of Endoskeleton—Connective Tissue—Cartilage—Bone—Axial Skeleton—Appendicular Skeleton—The Limbs—Pectoral and Pelvic Arches. Physiology—Alimentation—Respiration—Circulation—Epidermis—Epithelium—Muscular Tissue—Muscular System—Nervous System—Spinal Cord—Nerve Trunks—Brain or Encephalon—Cerebellum—Cerebral Hemispheres—Ventricles of Brain—Cerebral Nerves—Sympathetic System—Organs of Special Sensation (Smell, Sight, Hearing)—Reproduction—Tadpole—Distinctive Characters of Vertebral Animals, - 133

GLOSSARY, ... 167

INDEX, ... 191

GENERAL BIOLOGY.

CHAPTER I.

1. BIOLOGY is the science which treats of *living beings*, whether animal or vegetable.

2. In order to understand properly the nature, effects, and conditions of life, it is best to begin with the simplest forms in which it manifests itself, and wherein the attendant circumstances present the fewest possible complications. Our first example will be

THE TORULA.

3. **Yeast** or **barm**, by filtration, may be separated into a solid and a liquid portion. The latter, when examined with a microscope, is found to consist of small spheroidal bodies (*corpuscles*), each an independent living organism, called a **torula**, floating in a fluid from which they imbibe their nourishment, and which, from this circumstance, is called the **plasma**. The latter is a general term signifying the elaborated fluid from which living bodies immediately derive their substance. The best plasma known for the yeast plant is a preparation which goes under the name of Pasteur's fluid. It is constituted as follows:—

Water,	8376 parts.
Cane sugar,	1500 ,,
Ammonium tartrate,	100 ,,
Potassium phosphate,	20 ,,
Calcium phosphate,	2 ,,
Magnesium sulphate,	2 ,,
Total	10,000 ,

The torulæ are minute spheroidal bodies, whose existence in the saccharine solution is essential to the production of yeast. The same fluid may also hold a number of little rod-like bodies, called *bacteria*, of which more will be said in a future chapter.

4. Living beings may be regarded under two distinct aspects:—

(*a.*) The **Morphological** aspect. By *morphology* is meant the form, structure, and molecular arrangement of a living body.

(*b.*) The **Physiological** aspect. Under the term physiology is comprehended all the *activities* or *functions* of the organism. Every example coming under our notice in these pages shall be examined under both these aspects.

5. **Morphology of Torula.**—To the naked eye, yeast looks like a muddy fluid. But under the microscope, the corpuscles found floating in it have a definite spherical or ellipsoidal form. They vary from $\frac{1}{2500}$ to $\frac{1}{7000}$ of an inch in diameter; but their general average size is $\frac{1}{3000}$ of an inch, or about the size of the red corpuscles of the human blood.

Torulæ are scattered loosely through the plasma, or

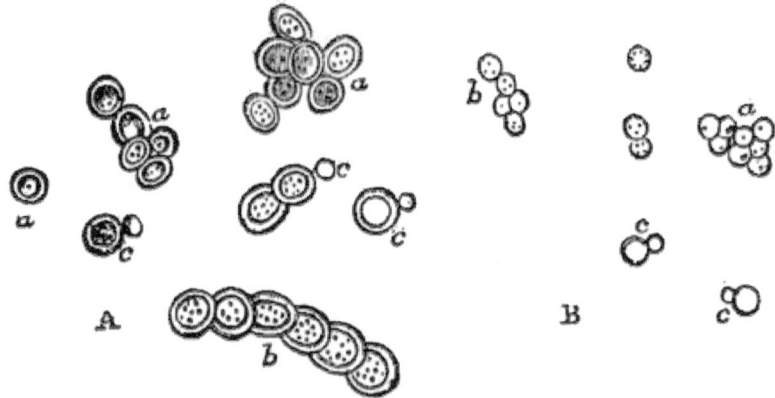

Fig. 1.—A, Torulæ, single and in groups; *c*, Torulæ seen budding; all highly magnified. B, The same, not magnified so much.

oftener they are associated in groups. The forms of these groups vary considerably, but their general appearance may be indicated by fig. 1, A, B.

6. On examining a single corpuscle, we find it to be a *cell* (fig. 2), and consisting of the following parts:—

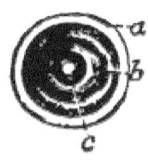 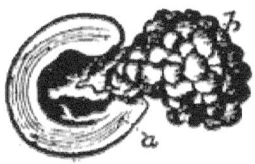

Fig. 2.—Diagram of Torula; *a*, sac; *b*, protoplasm; *c*, vacuole.

Fig. 3.—*a*, Sac, burst; *b*, protoplasm flowing out.

(*a*.) An outside transparent ring.

(*b*.) A semi-fluid matter contained in this, having the appearance of ground glass.

(*c*.) Sometimes a small space inside, more clear than the rest.

By pressure, a fracture may be made in the outer layer, or investing envelope, and the cell contents will flow out, as represented in fig. 3. We thus observe that the corpuscle consists of—

1. A structureless sac, or *cell wall*, containing
2. a semi-fluid substance, the *protoplasm*.

The interior of the protoplasm is more fluid than the rest, and is therefore more transparent. This central space is sometimes called a *vacuole*. The protoplasm itself may thus be regarded simply as a hollow closed sac, but the latter term is employed exclusively to denote the cell wall.

7. When the torula is treated with a solution of iodine or magenta, the protoplasm becomes stained, while the cell wall remains unaltered. When dilute caustic potash is employed as a re-agent, the protoplasm is dissolved out, and the sac or cell wall remains. By this means, as also by the use of acetic acid, they may be separated to be examined, and shown to be *of different chemical com-*

position. The *sac* is composed of *cellulose,* a substance consisting of carbon, hydrogen, and oxygen, the two latter elements being in the proportion in which they form water. It is almost of the same composition as starch, gum, and woody fibre. The latter, indeed, may, by being digested with weak sulphuric acid, be turned into sugar; and brandy is made in America from sugar thus manufactured. Cellulose is susceptible of a similar transformation. The wall sac, however, of the torula, in addition to cellulose, contains also water and mineral salts.

8. The protoplasm is more complex. In general, it may be said to consist of

>Protein compounds,
>Fat,
>Mineral salts, and
>Water.

9. **Protein.**—By dissolving the protoplasm in dilute caustic potash, and then acting upon it with acetic acid, *casein* will be formed. This substance is analogous to the *fibrin* of animals, and the *gluten* of vegetables; and, like them, is a protein compound containing the four elements, carbon, hydrogen, oxygen, and *nitrogen*, with some sulphur. The exact proportion is—

Carbon, . . . $55\frac{1}{2}$
Hydrogen, . . . $7\frac{1}{2}$ } *Plus* some sulphur.
Oxygen, . . . 23
Nitrogen, . . . 14

Fat consists of carbon, hydrogen, and oxygen, but in this instance there is *more* hydrogen than would form water with the oxygen. The *mineral salts* are six per cent. of the weight of the cell, principally salts of potassium. In this particular, the protoplasm of the torula differs much from the cellulose, as the latter contains but a small quantity of mineral matter. It also differs from the cellulose in containing no starch.

10. The destructive analysis of yeast by the older chemists yielded the same inorganic constituents as animal

matter, principally carbonic acid, water, and ammonia. Modern researches have determined that the mode of development and growth of the cell is just the same as is observed in the elementary tissues of animals and plants.

11. Physiology of Torula.—The torula floating in its plasma will *grow* if the proper temperature be maintained.

The sustenance and growth of the torula take place by

(*a.*) The imbibition of the plasma through the cell wall.

(*b.*) Its subsequent elaboration into "formative material," or "blastema," through the influence exerted upon it by the active living principle which resides in the protoplasm;

(*c.*) The *assimilation* of the "organisable blastema," as it is taken up by the protoplasm to form part of its own living substance.

The increase in size of the cell wall takes place by the process called *intussusception* (laying hold of inwardly), *i.e.*, by the assumption of new molecules interstitially deposited among those previously in existence, and derived immediately from the protoplasm by the metamorphosis of its own substance. It is important to observe that its growth is not a result of the formation of new layers externally, nor of the continued extension of the outer layers by new matters laid up internally.

12. Reproduction in the torula takes place by *gemmation*, or *budding*. This consists in new individual corpuscles growing out of the parent cell in the form of *buds* or outgrowths. These remain for some time attached to the parent, and continue to grow more or less independently of it. When nearly full grown, they either drop off, or remain attached to the parent cell, and to each other, forming groups (fig. 4). If even a single torula be deposited in Pasteur's fluid, the liquid, previously clear, will soon become muddy like yeast, by the development and growth of multitudes of torulæ.

Multiplication also takes place by the breaking up of the protoplasm into a large number of minute cells, which, by their growth, finally burst the wall of the parent cell, and escape as young torulæ (fig. 5).

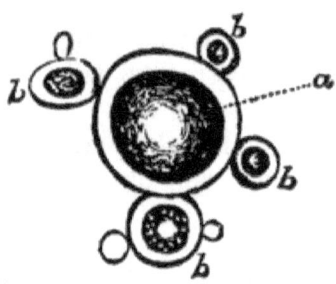 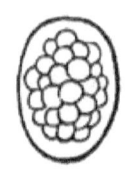

Fig. 4. DIAGRAM SHOWING GEMMATION IN TORULA. *a*, Parent cell; *b*, young cells in various stages of development.

Fig. 5.—TORULA, showing the protoplasm broken up into a mass of cells, which, after rupturing the investing sac, assume each an independent existence.

13. The torula introduced into Pasteur's fluid exercises a wonderful influence over its constituents; old chemical combinations are broken up, and new ones formed as a mere consequence of its existence in the fluid. It is through the mysterious power of the protoplasm that all this influence is exerted. Concurrently with these transformations in the plasma, other changes are going on in the cell itself; there is an absorption of oxygen, and a giving off of carbonic acid. This proves that the substance of the torula is wasting as well as growing.

14. The protoplasm must be of a particular form and constitution in order that the torula may absorb the nutritive materials from its plasma. It must also, of course, be supplied with the elements of protein, etc., in the plasma, else it could not elaborate the "formative material" required for its sustenance and growth. The temperature must not be over 60°C., nor under 0°C. Should it rise above 60°C., the torula will have its vitality destroyed; should it descend below the freezing point, the vital powers will be suspended. 70°F. or 21°C. is the most favourable temperature for the growth of the torula.

Light is not essential to its existence or growth. The phenomena of vital action in this case go on in the dark as well as in the light. This fact is important as bearing on biology; for it is well known that plants in general cannot have at least a healthy existence in the absence of light.

Pasteur has found further, that if sugar be present, the growth of the torula may take place without the aid of *free* oxygen.

15. It is of course to be expected that when the torula dies, its constituents have lost their stability of chemical equilibrium; that they are, therefore, resolved again into the same inorganic forms as animals and plants are reduced to when they die and decay, namely, carbonic acid, water, ammonia, and mineral salts.

16. In the process of fermentation, a fluid containing sugar is turned into one of quite a different character. Carbonic acid is given off; alcohol—a compound of carbon, hydrogen, and oxygen—remains. We should then expect (as was formerly believed to be actually the case), that the weight of the carbonic acid, together with that of the alcohol, ought to be exactly the weight of the sugar. The fact, however, is that only 95 per cent. of the weight of the sugar is thus accounted for. Two other substances, viz., glycerine and succinic acid, both compounds of carbon, hydrogen, and oxygen, are also to be found in the fermented sugar; but they only amount to 4 per cent. of the whole, leaving 1 per cent. of the sugar still to be accounted for.

100 parts of sugar (after fermentation) contain of

Alcohol + carbonic acid, . . .	95
Glycerine + succinic acid, . . .	4
Total, . .	99

17. Fermentation can only take place in a solution of sugar, in which torulæ are growing and multiplying. If no torulæ be admitted, the liquid will never ferment. But if a single torula be introduced, fermentation imme-

diately sets in. Hence fermentation *is the result of the torula's vitality.* Torulæ may exist in a fluid containing no sugar in solution, but then ordinary fermentation, or the disengagement of carbonic acid, cannot take place. No good explanation has yet been given of the cause why fermentation must go on when the torulæ exist in the saccharine solution. Two hypotheses have, however, been advanced to account for it.

(*a.*) The first theory sets forth that the torula, by feeding upon some one of the constituents, thus destroys, as it were, the chemical balance of the remainder. These must, in consequence, re-arrange themselves, but in a manner totally different, to restore the equilibrium.

(*b.*) The second hypothesis supposes that the very *vitality* of the torula tends to destroy this equilibrium by giving an impulse to the decomposition of the molecules in contact with it, just as the *heat* of one's hand is sometimes sufficient to decompose certain chemical compounds. It is not necessary, of course, to suppose that it is the vital heat of the torula that produces the effect referred to, though we know that heat, and light also, as "modes of motion," have the power of shaking asunder, so to speak, the atoms of which molecules are composed. Fermentation is then simply a *collateral effect* of the presence of the torula, but is not essential to its existence. This latter is the more probable hypothesis.

CHAPTER II.

PROTOCOCCUS.

18. If the green scum which is often observed collecting on the roofs of houses and other such situations, be mixed with water, and examined under a microscope, it will be found to consist mainly of small bodies (each a protococcus), between $\frac{1}{10,000}$ and $\frac{1}{380}$ of an inch in diameter. They are coloured, some strong green, some strong red, some half and half. The rain water in which they are constantly bathed may be regarded as the plasma.

19. Rain water is not absolutely pure. It contains salts of ammonia (nitrate, carbonate, etc.), as well as carbonic acid and oxygen in the free state. Dust also finds its way into it. Dust contains salts of potash and lime, with other mineral matters. All these different materials in solution make up the plasma of the protococcus.

20. **Morphology of Protococcus.**—In a general way, the protococcus resembles the torula. It is found under two different forms, the still or quiescent, and the mobile. In the quiescent form, there is a sac of cellulose surrounding the protoplasm. The latter is filled with granules of a green or red colour, which communicate their tint to the plasma regarded as a whole. The cell wall is quite transparent. If the latter be subjected to pressure, it will burst, and the protoplasm will fall out as a semi-fluid mass (fig. 3). The green and red colours are due to the particles of chlorophyll contained in it, which may, contrary to the derivation of the term, be either *red* or *green*. The torula has no chlorophyll in its protoplasm.

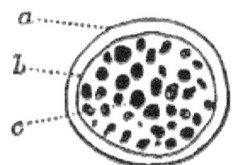

Fig. 6. A Cell of Protococcus. *a*, Sac; *b*, protoplasm; *c*, chlorophyll granules.

21. **Mobile Forms.**—These may be seen swimming through the liquid, generally without any sac. They tumble, rotate, and pull or haul. They move by cilia. Occasionally they are observed surrounded with a sac, and sometimes the sac appears at a considerable distance from the protoplasm (fig. 7, *b*, *c*).

The mobile form is sometimes seen to drop to the bottom, and assume the quiescent form.

22. Magenta stains the protoplasm of a very dark

colour. Caustic potash will dissolve out the protoplasm, and leave the sac untouched.

Fig. 7.—Groups of Protococci (highly magnified), showing comparative sizes and fission, *a* ; *b*, mobile form without sac ; *c*, mobile form with sac; *d*, cilia; *e*, groups of very small protococci.

Iodine and sulphuric acid (both together) will turn the wall sac blue. This is not the case where they act on the torula; but when cellulose is *pure*, which is not the case in the latter instance, this change will always take place.

$$\text{Sac} = \begin{cases} \text{Water.} \\ \text{Cellulose.} \\ \text{Mineral salts.} \end{cases} \quad \text{Protoplasm} = \begin{cases} \text{Protein.} \\ \text{Water.} \\ \text{Fat.} \\ \text{Mineral matter.} \end{cases}$$

It is not quite certain whether the protoplasm contains starch.

23. Physiology of Protococcus—Growth and development.—A protococcus of the quiescent form placed in rain water, and exposed to sunshine, first enlarges its size. The mode of growth is the same as in the torula, that is, by the process of intussusception through the vital power of the protoplasm.

Multiplication takes place by fission. This is the commonest method of cell multiplication both in the animal and vegetable kingdoms. The division is into 2, 4, 8 portions, or generally some power of 2. This mode of multiplication (by fission), is also found in the mobile form of the protococcus. By a sort of hour-glass constriction, the protoplasm first separates into two parts, each of which takes a separate wall sac of cellulose.

These parts again subdivide in a manner perfectly similar. The segments may cohere for some time, but this cohesion is only temporary; ultimately, each becomes an independent individual. Propagation by budding is also observed, but rarely.

24. Sometimes the protococcus is found growing on the snows of the Arctic regions, and on the snow-capped summits of the Alps, where it forms patches called "red snow" (*Protococcus Nivalis.*)

25. The protococcus exactly *resembles* the torula in its ability to break up the materials of the plasma, and to re-arrange them in such new combinations as will serve the purpose of nutrition; but the materials from which the protococcus derives its nourishment are more crude.

26. The protococcus, however, *differs* from the torula in its power of breaking up carbonic acid, and evolving oxygen, retaining the carbon. That the former *does* act in this fashion, may be proved by placing some of those organisms under a glass jar containing carbonic acid, and exposing it to sunlight. After some time it will be found that the carbonic acid contained in the jar has been replaced by oxygen. It is only under *bright* sun light that this absorption of carbon, and evolution of oxygen, can take place. The presence of chlorophyll is also necessary, whether red or green. The protococcus, in the presence of nitrate of ammonia, free carbonic acid, and certain mineral salts, can add to its substance, and maintain its own existence by simply breaking up those inorganic substances into their constituents, and re-arranging them into protein; this the torula cannot do. The distinction looks like a contradiction in the action of the two protoplasms.

27. In the absence of sun light, however, *all* plants become oxidised, and evolve carbonic acid. But the chlorophyll is a superadded instrument in the protococcus; it acts under the light like a laboratory in turning the carbonic acid into a material for *food*, or performs, rather, the functions of a stomach in digesting the carbonic acid which it eats as food, than of a respiratory apparatus to aid

in oxidation. There is in reality no contradiction, as the consumption of carbonic acid in the protococcus is simply a nutritive operation.

28. **Mobile Protococcus—Cilia.**—The vibratile movements of the cilium are due to the alternate contraction of the opposite sides. In this instance, then, the protococcus has a power of contractility not observable in the torula; not only do the *cilia* contract thus, but the *protoplasm* also. This mobile form, sooner or later, looses its cilia, and takes a sac of cellulose, becoming a quiescent protococcus. The protoplasm is the original living substance, the cellulose being derived from it. It may be asked, how is the cellulose derived from the protoplasm? Does the substance of the cell wall *exude* through the protoplasm from the interior, or is it elaborated on the outside of the protoplasm? This latter is the more probable hypothesis, inasmuch as the cell wall itself contains a small quantity of nitrogen.

29. The element essential to the life of the organism is the protoplasm, and whatever is further elaborated beyond this must arise from the metamorphosis of the protoplasm. The cellulose, indeed, is not dead, for it is impregnated with the protoplasmic fluid. The thickness of the cell is increased by the intussusception among its older molecules of new materials elaborated by the protoplasm, and anything capable of growing in this way must have some degree of vitality.

CHAPTER III.

THE AMŒBA (A PROTOZOON).

30. **Amœba.**—This little organism lives in mud, or rather in the water which pervades it. The water serves as the plasma of the protozoon. There is some difference between the nature of this plasma, and that of the torula

and protococcus. In the two latter, the plasma consists of merely inorganic compounds, not containing protein ready formed, but certain constituents which the organism is capable of *converting* into protein, and the other substances necessary to its existence. The plasma of the amœba must, however, contain *protein ready made* in some one or other form. In fact, the food of the amœba may even exist in the plasma in the *solid* condition, out of which the organism is able to elaborate the materials necessary to its existence.

31. Morphology of the Amœba.—The form of the amœba (fig. 8) is for ever changing, and hence the name.

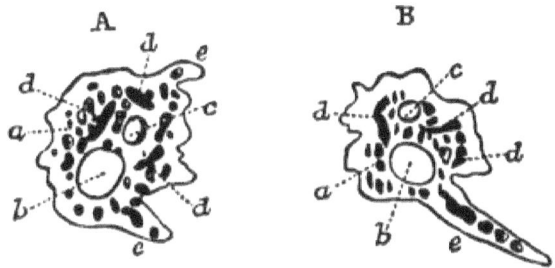

Fig. 8. A, B.—AMŒBA CHANGING ITS SHAPE. *a*, Protoplasm; *b*, nucleus; *c*, vacuole; *d*, particles of solid matter taken in as food; *e*, pseudopodia (greatly magnified).

There is no membranous sac as in protococcus and torula, but the body is a mass of transparent gelatinous matter, containing granules, shells of diatomaceæ, etc. It has a *nucleus b*, and occasionally a second contractile space *c*, called a "vacuole," which exhibits a tolerably regular diastole and systole. The contractile space seems to be filled with water from the exterior. The granular matter is not close to the margin; there is no sac, but the inner part of the protoplasm is more fluid than the rest. When a *pseudopodium* projects outwards, as at *e* (fig. 8), the fluid flows into the projecting part following the outline.

The nucleus is simply a part of the protoplasmic substance offering a greater resistance. The amœba, in itself, corresponds to the *protoplasm* in torula and protococcus.

Motion is effected, not by cilia, but by extending the pseudopodia, or finger-like projections of the body substance, and drawing itself after them.

32. The amœba may, however, become *encysted*, that is, enclosed in a coat of chitin, similar to the leathery substance forming the covering of insects. The *nucleus* in the encysted form remains for some time, but the other matters in the protoplasm disappear.

33. If the amœba be impregnated with magenta, it is observed that the protoplasm will be coloured. It contains a protein matter analogous to that of the protococcus and torula. It also contains mineral matters, such as salts of potash.

34. **Physiology of the Amœba.**—The amœba multiplies by *fission* when in the encysted form. The nucleus divides first, afterwards the protoplasmic mass.

35. The torula *absorbs* the plasma, and *manufactures* from it protein in a condition fitted for at once becoming part of the substance of the organism. The amœba cannot do this; there must be suspended in the plasma solid matters *containing protein,* which the amœba seizes hold of with its pseudopodia, and draws into its substance. If we examine the interior of an amœba under a microscope, we shall discover protococci, etc., moving about

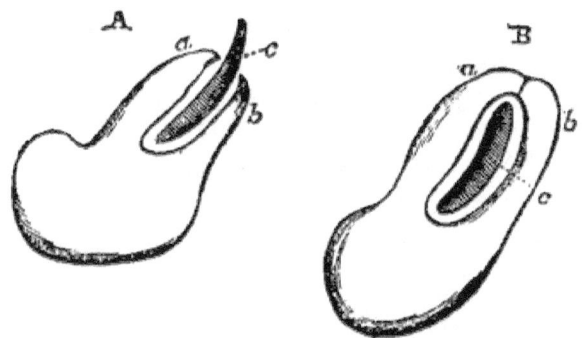

Fig. 9.—Diagram Representing Amœba Swallowing a Diatom. A, With pseudopodia (*a, b*) seizing a diatom, *c;* B, with pseudopodia (*a, b*) coalescing after drawing in the diatom.

in the fluid, and undergoing digestion. A and B (fig. 9) show how the amœba seizes on the solid matters of its food; *c* is a diatom round which, as shown at A, the amœba throws its pseudopodia. At B, the pseudopodia are represented as thrown still further outwards, until they meet and coalesce, thus bringing the diatom at once into the body of the organism, where it keeps floating about until the nutritive part is consumed; then the innutritious portion finds its way out *anywhere* through the gelatinous matter. In this fashion the amœba improvises for itself both a mouth and an anus. It thus differs widely both from the torula and the protococcus, inasmuch as it does not *manufacture* protein, but lays hold of such as may already be within reach.

36. Constant oxidation is necessary to the existence of the amœba. It is necessary that the water in which it lives should contain free oxygen. If this be removed from the water, the animal dies. The absorption of oxygen, and the evolution of carbonic acid are just as necessary to it as to any of the higher animals.

37. It also gives off a nitrogenous compound, which proves that it is constantly *wasting* protein. It cannot *make* protein out of inorganic elements; it *may* be able to make fat, but it is essentially a *waster* of protein. On the other hand, the torula and protococcus are *manufacturers* of protein. As to what is the ultimate fate of an individual amœba, the probabilities are that it lives its life, and dies like other animals, unless we suppose that it becomes revivified in the process of multiplication.

38. Electricity stimulates the contractility of the amœba. Electric shocks will cause it to contract its pseudopodia. This fact is important as bearing on the contractile force of the shock on the muscles of animals.

39. The conditions of existence are:—
 1. A supply of protein matter.
 2. A temperature not lower than 0° C., and not higher than 40° or 45° C.

At the former temperature, the vital powers become

dormant; at the other extreme, the animal dies, contractions cease to manifest themselves, and the substance of the mass undergoes coagulation.

40. The **Colourless Corpuscles** of the blood are little spheroidal bodies found floating in the blood of man, and, indeed, in that of all animals in which a circulating nutritive fluid has been observed. They are independent, active organisms, very like the amœba in the fact that they consist of a single cell, are constantly changing their shape, and throwing out pseudopodia (fig. 10). Each corpuscle is from $\frac{1}{3000}$ to $\frac{1}{2500}$ of an inch in diameter, possessed of a thin cell wall and provided with a nucleus. The latter also is colourless; but in some *rare cases* it has been observed slightly tinted with red. In healthy human blood, these colourless corpuscles are much less numerous than the red corpuscles, being only in the proportion of 1 to 300; but in all invertebrate animals, and even in the earliest stages of embryonic development of vertebrate animals, the blood has only colourless corpuscles. They are very abundant in inflammatory blood, and they predominate in that morbid condition of it known as *Leucocythaemia*. The circulating fluid of the lymphatic system contains no red corpuscles, but colourless corpuscles abound in it.

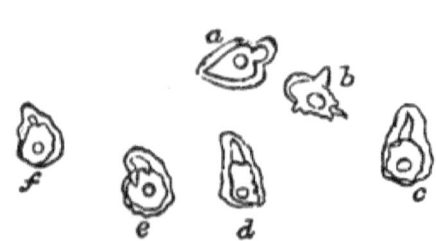

Fig. 10.—*a, b, c, d, e, f*, Successive forms assumed by the colourless corpuscles in the blood of a newt in the course of a few minutes (greatly magnified).

The origin and functions of the colourless corpuscles have not yet been satisfactorily determined. In reference to the former, Professor Huxley observes, that "it is highly probable they are constituent cells of certain parts of the solid substance of the body which have been detached and carried into the blood, and that this process is chiefly effected in what are called the *ductless glands*, from whence the detached cells pass as lymph corpuscles, directly or indirectly, into the blood."

Respecting one, at least, of their supposed functions, the same eminent authority thus states his opinion: "There is very great reason for believing that the red corpuscle is simply the nucleus of the colourless corpuscle somewhat enlarged, flattened from side to side, changed by development within its interior of a red colouring matter, and set free by the bursting of the sac or wall of the colourless corpuscle. In other words, the red corpuscle is a free nucleus." *

41. COMPARATIVE VIEW OF THE MORPHOLOGY AND PHYSIOLOGY OF THE THREE UNICELLULAR ORGANISMS:—

TORULA.	PROTOCOCCUS.	AMŒBA (*Protozoon*).
	STRUCTURE (Morphology).	
Sac	Sac	......
Protoplasm	Protoplasm	Protoplasm.
......	Cilia	Pseudopodia.
	CHEMICAL COMPOSITION.	
Protein—Sulphur	Protein	Protein.
Cellulose	Cellulose	Glycogen (?)
......	Starch (?)	......
......	Chlorophyll	......
Fats	Fats	Fats.
Salts	Salts	Salts.
Water	Water	Water.
	PHYSIOLOGY.	
......	Contractility.	Contractility.
Constructs Protein.	*Constructs* Protein.	*Ingestion* of Protein.
Multiplication by Gemmation.	Multiplication by Fission.	Multiplication by Fission.

42. We thus observe in comparing these three typical forms, the torula, the amœba, and the protococcus, that protoplasm is essential in all three; and that the fundamental constituent of the protoplasm is protein, to which we find added in each, fats, water, and mineral salts.

* *Lessons in Elementary Physiology*, p. 60.

Of the three, the torula is the only one that exhibits no evidence of contractility.

It is not certain whether the protoplasm of the amœba contains glycogen, a substance corresponding to cellulose or starch; but, in the encysted form, it is enveloped in *chitin*, a material which, though it contains nitrogen, is somewhat allied to cellulose.

43. It has been already observed that the torula and protococcus *manufacture* protein, while the amœba *destroys* it. Of these three organisms, the torula is the type of the fungous tribe of vegetables; the protococcus, that of the green vegetables, which contain chlorophyll; and the amœba is the type of the animal kingdom, for it can be shown that all animal tissues owe their origin to cells exactly like the amœba in all essential particulars. All living beings may therefore be conveniently grouped in three great divisions, viz., *Fungi*, *Plants*, and *Animals*.

CHAPTER IV.

BACTERIUM.

44. When we make an infusion of hay in *warm*, or even in cold water, allow it to remain for some time, and then filter it to take away all the solid matter, we have what is called hay tea. At first the fluid appears clear and tranquil. If left to stand for thirty-six hours, it becomes turbid; a scum is then observable, which gradually becomes denser and stronger. In a week or so, the aromatic smell passes away. If examined under the microscope, the liquid seems full of rod-like bodies, called **bacteria**, whose diameter is only $\frac{1}{20,000}$ of an inch. The length varies from twice to fifty or sixty times the breadth. They seem to be obscurely jointed transversely, like striped muscular fibre. It is supposed that each segment is simply one of the small bodies *c*, fig. 11, united to its neighbour by a very thin partition. Iodine colours the central part, leaving the sac stainless. Each coloured dot is a protoplasm, the adjoining cells being held together by a transparent substance forming the cell-sacs.

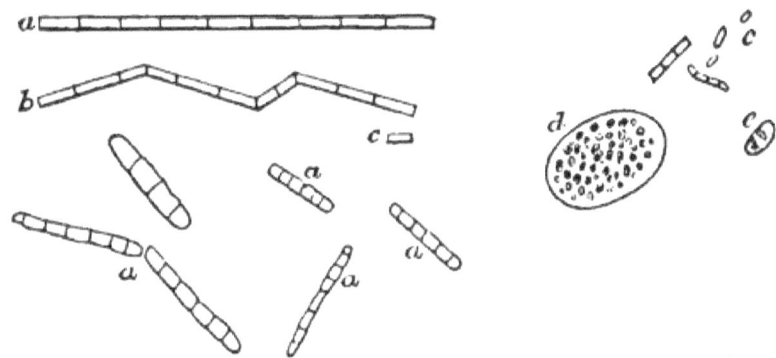

Fig. 11.—Appearances Presented by Bacteria under the Microscope. At *c*, are isolated bacteria; at *d*, they are arranged round a centre; while at *a* they appear in long strings; at *e* is observed a solitary torula. All highly magnified.

45. There are several varieties of the bacterium, but they are all alike remarkable for their power of independent motion. Their movements are somewhat peculiar, and vary according to the form of the organism. The *oscillariæ* (or *oscillatoriæ*) have an oscillating motion, often excessively rapid, round a fixed point, somewhat resembling the movements of a toy serpent. The *vibrions* show a wave-like motion in the same plane, like that of serpents through a field of grass. Others, as the *spiræolæ*, move in the line of a spiral or cork-screw.

46. But these minute organisms also exhibit movements when dead, which must be carefully distinguished from those that are entirely due to their vitality. It is highly important not to confound the two kinds of motion. That of dead matter is merely mechanical, not vital, and resembles somewhat the swinging motion given to an hour glass, or that of "a floating buoy round its mooring." Finely divided camphor, cinnabar, or gam-

boge, will exhibit movements of a similar character. The phenomenon is called the *Brownian movement*, after Brown, who thoroughly investigated it. It is simply a molecular movement due to mechanical, not to vital causes.

47. The bacterium, when about to multiply, settles down into the quiescent form, becomes *encysted, i.e.*, takes a covering of gelatinous matter, which becomes more or less encrusted on the outside. The original protoplasmic substance breaks up into granules, which ultimately escape from their covering, and begin to germinate, throwing out fine filamentous bodies, each called a *leptothrix*.

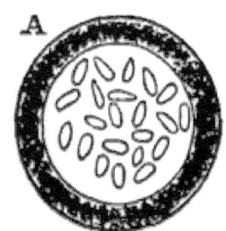 

Fig. 12.—A, Bacterium encysted. B, The granules after escaping from the cyst.

48. Putrefaction is a collateral phenomenon attendant upon the vital action of the bacterium, just as fermentation is an accompaniment of the living activity of the torula, though not essential to its existence. Torulæ are found accompanying the bacterium, but it is questionable whether they be of the same kind as are found in yeast.

49. To show that putrefaction is the direct result of the vitality of the bacterium, the following experiment may be performed.

Take three vessels containing an infusion of hay, boil the second and third, but not the first; leave the second open, and close the third with a plug of cotton wool.

The infusion in the first will putrefy and exhibit all the phenomena detailed above, that in the second will do the same, but after a longer interval, while the infusion in the third vessel will remain perfectly sound for six months—nay, for six years if the proper precautions be taken. In the first, the bacteria are killed; in the second they are ~~also~~ killed, but the vessel lies open for the admission of spores; in the third, the spores are prevented from getting in.

50. The bacteria, or their germs, exist in the hay originally, as do also infusoria, rotifers, etc., and hence the hay infusion must also contain the bacteria. They may likewise exist in the air, and hence may find their way into the fluid.

51. It has been proved by experiment that if the vessel be kept *chemically clean*, no bacteria will be formed in a boiled infusion, even though not protected by a plug of cotton wool. The explanation seems to be, that the germs of the bacteria are so light that they can never *descend* in still air as far as the fluid. But in any case, the plug of cotton wool will be an effectual stopper. Experiments such as these seem conclusive as to the question of spontaneous generation, proving, as they appear to do, that every living organism must be derived from a germ. It must be admitted, however, that many very eminent naturalists hold a contrary opinion, and they even appeal to the results of experiments to justify their conclusions.

CHAPTER V.

THE PENICILLIUM.

52. **The Penicillium** is a fungus, the ordinary "green mould" which grows on decaying animal or vegetable matter. It would grow in Pasteur's fluid without the sugar, inasmuch as that fluid, so modified, contains all the ingredients necessary to its existence. Under favourable circumstances it grows in great quantities. It is formed of filaments so closely interlaced with each other

as to appear like a sort of felt-work or paper. This papery matter is the mycelium. On the surface there is observed a large quantity of a greyish white powder. This consists of *conidia*, each conidium being about $\frac{1}{3000}$ of an inch in diameter. From the mycelium filaments hang down in the liquid like silken fibres. These are called *hyphæ*, and are the essential organs of the penicillium. Each hypha (fig. 13) consists of a string of cells, arranged end to end like the beads of a necklace. Each segment, again, constitutes one cell, like a torula so far as its essentials are concerned. Iodine or magenta stains the protoplasm, but not the sac. With caustic potash the protoplasm dissolves out, leaving the sac behind. Very often the hypha is bent and branching. It has no longitudinal divisions; they are all transverse. If we supposed the torulæ, when coherent, to attach themselves longitudinally, we should then have a hypha instead of a cluster of cells

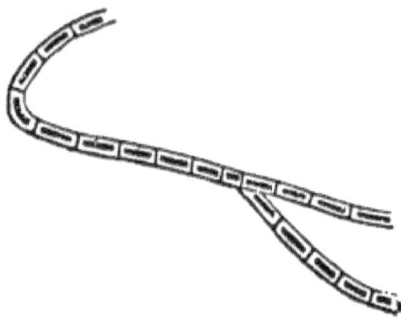

Fig. 13.—A Hypha, bent and branching.

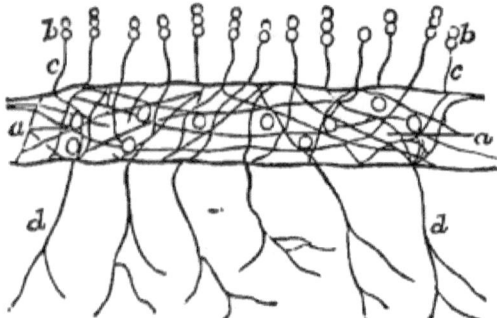

Fig. 14.—Diagrammatic Representation of Penicillium. *a*, Mycelium made up of interlaced hyphæ, and interspersed with conidia ; *b*, conidia ; *c*, conidiophores ; *b* with *c* constitute subaërial hyphæ ; *d*, loose hyphæ suspended from the mycelium.

such as is observable in torulæ. The *mycelium* is formed simply by the interlacing of the hyphæ (fig. 14, *a*), and is therefore capable of maintaining its own vitality. The mycelium never constitutes a hard mass, but merely a sort of loose felt-work.

53. There are also aërial hyphæ (fig. 15; see also fig. 14, *b*, *c*). They, however, have the power of covering themselves with a sort of wax, which always keeps them dry. Their special designation is "conidiophores." At the upper end they present the appearance of a *brush*, and hence the name "penicillium." The cells as they approach the ends become less and less attached to each other. Ultimately they fall off, each cell becoming a conidium or spore, and acting as an independent organism like a torula. It thus increases itself by multiplication, and becomes in time a full grown penicillium. The conidia fall into the interstices of the mycelium, and by growing become interlaced with each other, forming a new mycelium, or adding to the fabric of the old one (figs. 16 and 17).

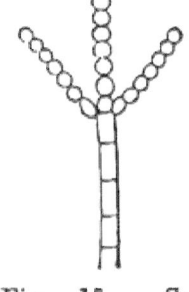

Fig. 15. — Sub-aerial Hypha, or Conidiophore, bearing conidia.

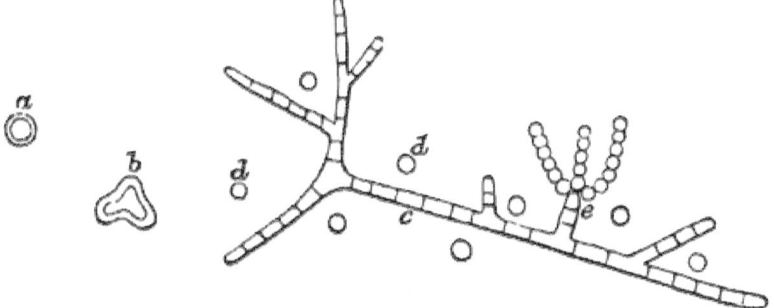

Fig. 16.—*a*, *b*, *c*, Successive stages in the development of a conidium into a hypha; *d*, *d*, are conidia which, by their subsequent growth, will form new hyphæ interlacing with *c*, and with each other; *e* is a conidiophore, which continually drops its conidia among the mycelium thus formed.

54. Conditions of Existence.—These are the same as in the torula. The nutriment is derived from organic matter—whether animal or vegetable—which will readily decompose, or which is already in a state of decay. In either case *the presence of the fungus hastens decomposition.* The penicillium thrives best in a damp and dark situation, because, firstly, moisture promotes the decomposition of the organic matters on which it subsists; and secondly, because the evolution of carbonic acid—a constant accompaniment of its existence—is best carried on in the absence of light. It has not been ascertained whether the penicillium requires free oxygen, but it absorbs it largely in whatever condition it may be presented to it, and gives off carbonic acid abundantly.

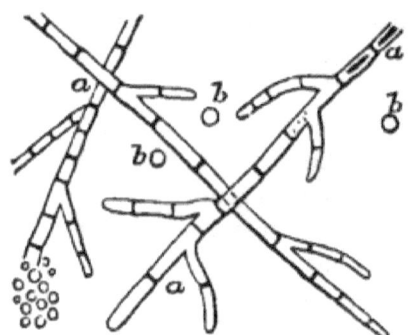

Fig. 17.—Hyphæ, *a*, in mycelium, as seen under the microscope, interlaced and branching, and interspersed with conidia, *b*.

55. Reproduction.—The penicillium is derived from a conidium. The conidium is a kind of spore. If put into a fluid in which it can live, it throws out hyphæ. Torulæ may or may not be developed. Becoming interlaced, the hyphæ form an expansion called a mycelium. This again gives rise to aërial hyphæ, differing from ordinary hyphæ in having their segmentation more thorough, and so giving rise to conidia. In all this development there is nothing but cell growth, and multiplication by fission (asexual reproduction), the cells being just like elongated torulæ, and like them, too, having no nucleus. There is, however, a stage wanting in the history of the penicillium. It has not yet been determined whether *gamogenesis*, or reproduction by the *union of two distinct protoplasms*, is one of its ordinary vital phenomena. Our observations upon this head, therefore, must be transferred

REPRODUCTION OF THE PENICILLIUM. 33

to a different example, the *peronospore*, in which such a mode of reproduction has been detected.

56. This fungus (*peronospora infestans*) is very like the penicillium. It is the cause of the "blight" in the potato. The mycelium is, however, quite loose—not nearly so much compacted as in penicillium. The hyphæ do not form a feltwork, but ramify along the air passages and intercellular spaces of the potato (fig. 18). They give off projections *into* the cells of the potato, and thus feed upon their protoplasm, and rapidly develop. The peronospore multiplies like the penicillium, in giving off subaërial hyphæ, which in their turn give rise to the production of conidia. These are carried off by the wind or by insects, are sown upon the tissues of plants yet untouched, and so the disease spreads rapidly.

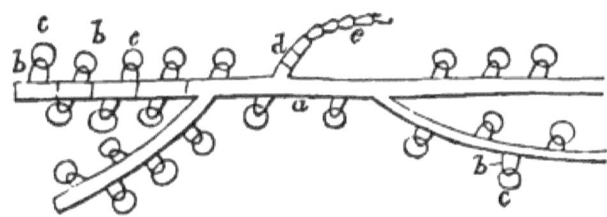

Fig. 18.—PERONOSPORA INFESTANS, the fungus which produces the blight in potato. *a*, Hypha branching; *b, b*, cellular projections piercing *c, c*, the cells of the potato; *d*, a subaërial hypha, with conidia, *e*.

57. After falling upon a congenial soil, the conidium begins to germinate. The protoplasm bursts through the sac, and being provided with cilia it keeps moving about for some time. It then settles down, envelops itself in a cellulose coat, and germinates anew like the penicillium. All this goes on during the summer months, while the potato plant is yet green.

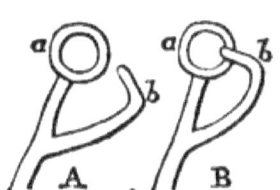

Fig. 19.—*a*, Oogonium; *b*, antheridium. In A, they approach; in B, they coalesce.

Towards autumn, however, certain parts of the hyphæ

produced by its germination assume a special activity. One part of it (fig. 19) develops a spherical mass (A, *a*) called the *oogonium*, while a neighbouring hypha, or a different part of the same hypha, develops a smaller mass *b*, designated an *antheridium*. When *a* becomes ripe, the antheridium *b* penetrates it. The protoplasmic matter of *b* enters into that of *a*, and becomes fused with it. The antheridium at once decays, but the oogonium becomes still more developed, surrounds itself with a thicker coat, and becomes an *oospore*. This is true sexual generation (*gamogenesis*), a mode of propagation which is very general both in the animal and vegetable kingdoms.

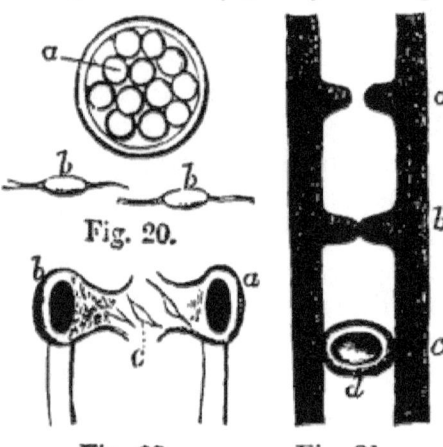

Fig. 20.—*a*, Segmentation of protoplasm in fertilised oogonium; *b, b,* ciliated spores.

Fig. 21.—CONJUGATION. *a*, Germ and sperm cells approaching; *b*, germ and sperm cells touching; *c*, coalescing to form a spore, *d*.

Fig. 22.—IMPREGNATION IN ACHLYA. *a*, Antheridium liberating antherozoids *c*, which pass into the spore sac *b*.

The protoplasm of the fertilised oogonium (fig. 20) divides and subdivides, and each segment ultimately (*i.e.*, during the next season) escapes from the envelope ciliated, and keeps moving about till it finds a place where it may germinate.

58. In some other forms there is even a simpler mode of sexual union, which goes by the name of *conjugation*. Two cells, A and B (fig. 21), strike out projections which approach each other, and even meet ultimately. The closed extremity of each projection is absorbed, the contents of both cells commingle, forming a new cell, *d*, which immediately becomes developed into a spore. In

the case of the peronospore, the antheridium, which contains the "sperm cell" or male element, is much smaller than the oogonium, which encloses the germ cell, or female element; where "conjugation" occurs they are of the same size, and as a new cell is formed by the fusion of the two protoplasms, it is difficult to decide which is the male and which the female element.

59. In the achlya the mode of cell union or "impregnation," as it is also called, is different from either of these two forms.

Two sacs are produced (fig. 22), one of which, a, contains the "germ cells," while the second, b, contains *antherozoids* or "sperm cells." When both become ripe the sacs open, and the ciliated antherozoids in b pass into a, and fertilise its contents. There may be something resembling this in the penicillium.

60. The penicillium is a type of the fungi. They all agree with it in certain particulars, in which they differ from all other vegetable forms.

Firstly: They take their origin in spores, either by hyphæ, or by division of the protoplasmic mass in the interior. They all give rise to spores, with or without "impregnation," or cell union. They begin to germinate in the same way as the penicillium, namely, by hyphæ with transverse segmentation. The hyphæ never divide longitudinally. They always have a filiform shape, and are all, however much interwoven, independent structures like the threads in cloth. The mushroom, for instance, though it seems to have a solid stem and cap, is nevertheless entirely formed of distinct hyphæ growing closely together. In many other fungi, also, the hyphæ become compactly interlaced.

Secondly: The gills (lamellæ) of mushrooms give off hyphæ, each bearing four conidia; each conidium, when sown, begins to germinate exactly as in penicillium (fig. 23).

Thirdly: All fungi *absorb oxygen*, and *give out carbonic acid*, just as animals do.

Fourthly: They are *wholly devoid of chlorophyll*, and hence cannot decompose carbonic acid.

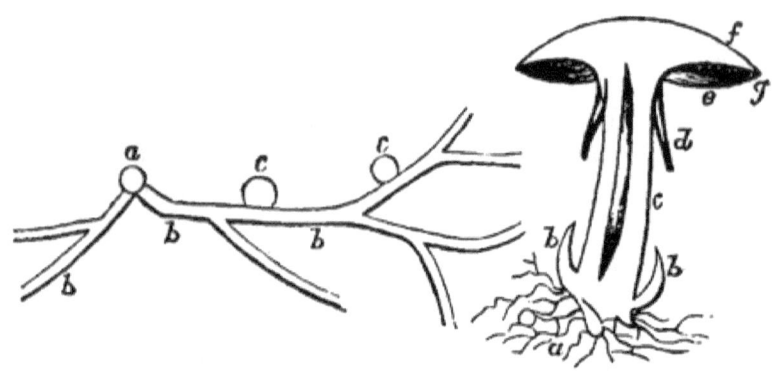

Fig. 23.—A, GERMINATION OF CONIDIUM OF MUSHROOM. *a*, Conidium, giving off hyphæ *b, b*, which branch to become a mycelium; *c, c*, incipient mushroom caps. B, Vertical section of a mushroom (*agaricus*). *a*, Mycelium (or spawn); *b, volva*; *c*, stipes or stalk; *d, annulus*; *e*, lamella of hymenium or "gills;" *f, pileus* or "cap."

It is thus observed that all the fungi are constructed upon the plan of the penicillium by the growth and interlacing of hyphæ. The hypha, again, is constructed after the manner of the torula, if we imagine the torula to multiply by transverse fission. The conditions of existence may be broadly stated to be the same in all.

61. The phenomena of reproduction in fungi may be distributed into—

1. Those in which multiplication takes place either by budding or fission (*vegetative, agamic,* or *asexual*).

2. Those in which the process of multiplication requires the fusion of two *distinct protoplasms* (*sexual*).

Of the latter there are three modifications:—

(1.) That in which the part representing the male element is smaller than the other, and where coalescence takes place by the "sperm cell" piercing the "germ cell" as in *peronospore.*

(2.) That in which both cells are undistinguishable in size or other appearances, and where the coalescence takes place in a new cell, formed by the union of other two.

(3.) That in which ciliated antherozoids enter an opening made for them in the sac containing the germ cells, as in *achlya*.

62. Green plants have their type in the protococcus; they all differ from fungi, in *possessing the element chlorophyll*, and in having thus the power of decomposing carbonic acid, and using part of it to build up their substance. Every green plant is made up of individual elements or cells, each like the protococcus in all essential particulars. Their structure, if examined, is found to be quite different from that of fungi. The fungi are formed by a *matting*, into a closer or looser fabric, of individual hyphæ, which are quite independent of each other. It is true that some of the lower green plants (confervæ for instance) have a structure more or less similar to this; but in all the higher green plants there is longitudinal as well as transverse division of the cells, and hence they are arranged in masses, as we find them in mosses and other green plants whose structure is wholly cellular. This is one great difference; a leaf-like expansion is not made up of elements like hyphæ, but by cell-division, longitudinally as well as transversely—in fact, in all directions.

CHAPTER VI.

THE CHARA.

63. THE **Chara** may be regarded as an aggregation of cells like protococcus. It possesses chlorophyll, and has the power, in consequence, of decomposing carbonic acid under sun-light, the carbon of which it retains to form part of its own substance. It is a long, filamentous plant, growing in water. It has rootlets springing here and there from the axis, but the main source of its nutriment is the water in which it lives. This water will be found to contain salts of ammonia, such as nitrate, also

carbonic acid, and oxygen in the free state. The branchlets (or *leaves*, as they are also called) are grouped in "whorls," *i.e.*, a number of them spring from the same point in the stem; the same arrangement being repeated at regular intervals along it (figs. 24, 25). The *main stem* is called the *axis*. The branch proper, where it exists, is a secondary stem, springing from the axilla of the whorl of leaves, the structural arrangement being a repetition of that observed in the axis. The "appendages" comprise the branches, the verticillate or whorled leaves, and the rootlets or rhizoids. The "leaves" also have lateral appendages, consisting of *antheridia, nucules* or *spore fruits*, and *leaflets* (figs. 25, 26).

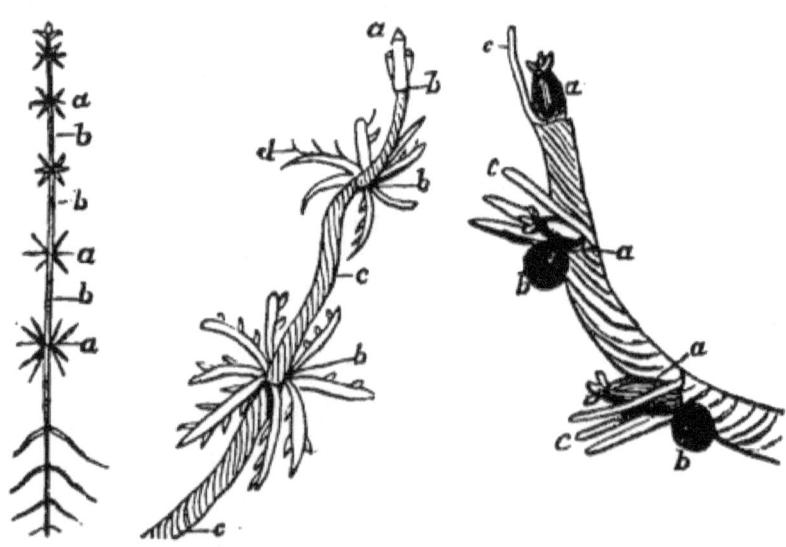

Fig. 24.   Fig. 25.   Fig. 26.

Fig. 24.—DIAGRAM OF CHARA. *a*, Node; *b*, internode.
Fig. 25.—CHARA VULGARIS. *a*, Summit; *b*, node; *c*, internode; *d*, whorl of leaves (or branchlets).
Fig. 26.—LEAF OF CHARA, magnified to show *a, a*, spore fruits; *b, b*, antheridia; *c*, leaflets.

Each *antheridium* is a little orange-coloured globule containing antherozoids. Contiguous to this, but farther up on the leaf, is a little bottle-shaped organ, somewhat larger than the antheridium, and called the *nucule* or spore-fruit. It contains the germ cell, thus constituting the oogonium, or ovary of the chara. The *leaflets* are small secondary appendages, one of which stands at either side of a nucule (fig. 26).

64. In examining the chara, we observe that the whorls of leaves are set along the axis at certain intervals, and that the stem is "segmented." This gives us an idea of what is understood when we speak of the homologies, or correspondencies of parts in the same structure. The *spore fruits* and *antheridia* are observed also to be homologous, and so are the rhizoids.

The whorls become closer as they approach the summit of the stem; the internodes become shorter, as do also the appendages (fig. 24). But all the internodes, sooner or later, grow to the full size of the under one, and there is, besides, a simultaneous development of new nodes and internodes. An internode is formed by the simple enlargement of a single cell, which continues to lengthen till it reaches its maximum size.

65. The chara "grows" at the summit alone; *i.e.*, no *new development* or evolution of cells takes place anywhere else in the stem. The lower cells simply add to their size; they *do not form new cells*. The apex is, therefore, called the "growing point" of the stem, and the plant itself is called an "acrogen" from this very circumstance. Ferns, mosses, and other cryptogamic plants grow nearly in the same way. The chara never thickens beyond a certain point, and this accounts for the slenderness of the stem.

66. Growth takes place at the apex by the development of new nodes and internodes. Each internode is seen to be formed by the growth and elongation of *one cell* (fig. 27). After macerating the stem in a weak acid, to remove any carbonate of lime with which it may be impregnated, it will be easy to examine the structure of each cell, and to observe that there are spiral striæ arranged round the wall of each internode (fig. 28). The

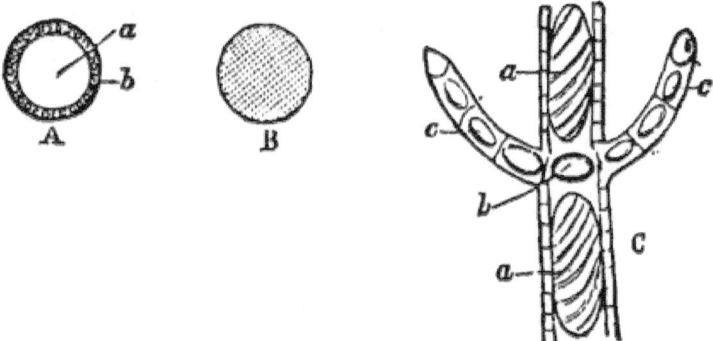

Fig. 27.—A, Transverse Section of Chara at Internode a, Medullary part; b, cortical part. B, Transverse section at node. C, Longitudinal section of chara. a, Internodal cell; b, nodal cell; c, cells of appendages.

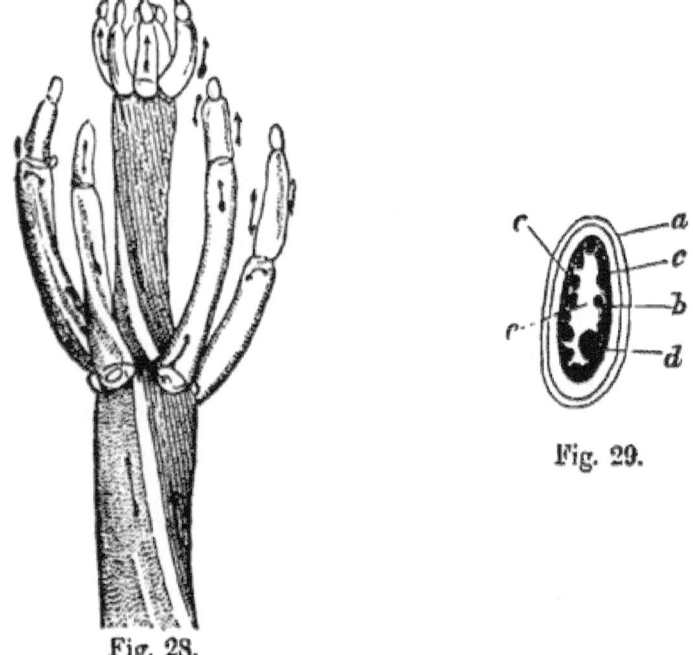

Fig. 28.

Fig. 29.

Fig. 28.—UPPER PART OF CHARA, to show the spiral striation, and the cyclosis. The arrows show the course of the currents.

Fig. 29.—INTERNODAL CELL IN CHARA. a, Cellulose sac; b, outer condensed layer of protoplasm, called the primordial utricle; c, c, granules of chlorophyll floating in protoplasm; d, nucleus; e, vacuole.

structure of the appendages is similar to that of the stem, but the mode of growth is best observed in the terminal bud of the latter. By the application of chromic acid or alcohol, the protoplasm may be separated from the cell wall; an internodal cell will then present somewhat the appearance set forth in fig. 29.

67. The chlorophyll is arranged in granules in the "primordial utricle" or outer layer of the protoplasm. The nucleus is not readily met with in fresh specimens, but chromic acid renders it apparent. The vacuole contains matter that may be regarded as accidental, or, at least, not essential. The granules of chlorophyll are quite free and unattached.

68. Beneath the chlorophyll granules there is, in each cell, a constant flow of protoplasmic matter, up one side of the cell and down the other. This circulation, called *cyclosis*, is owing to the contractile power of the protoplasm. Warmth favours this contractility, while electricity arrests it. The nucleus takes part in the rotation. Cyclosis in each cell is quite distinct from that observed in the cells adjoining (figs. 28 and 30).

69. The cells of the appendages do not, in reality, differ in their nature from those of the axis; but in the smaller cells the nucleus remains quiescent in the middle. Each cell of the chara is, in all essential particulars, exactly like the protococcus, except that the former possesses a nucleus, while the latter does not. The cells are also in their vital function independent of each other, and only so far modified that in their aggregation they make up the chara.

70. If we seek out carefully a terminal or growing bud, we shall find that it ends in a rounded surface, which is formed by a single cell (fig. 31, *a*). This cell, in the process of multiplication, divides by fission into two others, one of which is, so to speak, condemned to remain barren during its lifetime; while the other multiplies by the creation of new *lateral cells*, as well as by the duplicative multiplication which its parent underwent. These lateral cells, by their division in a similar duplicate fashion, give rise to the appendages; but in the latter case, when a certain stage has been reached, the terminal cell is found to be incapable of further division, and so the formation of new cells in that direction ceases. The lengthening of the leaf is effected by the simple enlargement of the cells already formed, without the production of any more cells. In the stem, however, this condition of the terminal bud is never reached; each cell as it is formed divides into two, one of which always remains single, while the other, besides giving off a new cell to continue the onward progress of the stem, develops also a circle of *lateral* cells to form the appendages.

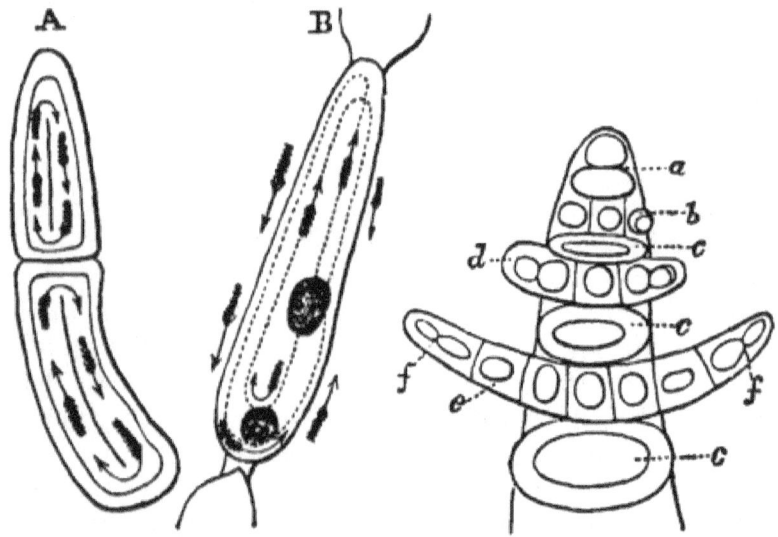

Fig. 30.—A, Cyclosis in Chara; the protoplasmic matter circulates in either cell, but does not pass into the other. B, Double current in a cell in the "hairs" of tradescantia. The *dotted* lines show a current in one direction, while the broken lines indicate a second current going in the opposite direction.

Fig. 31.—Growing Point of Chara. *a*, Terminal cell dividing; *b*, cells forming youngest node, and which by their fission will give rise to a whorl of appendages; *c, c*, internodal cells; *d*, incipient appendages; *e*, same farther advanced; *f, f,* terminal cell dividing.

MORPHOLOGY OF CHARA. 43

71. Side by side with the development of new cells, there is the simple growth or enlargement of those already formed. When the cell forming the first internode has reached a certain stage in its growth, its further progress is arrested. By this time, it is the longest of all, as the later ones are not yet fully grown. After some time, the second internode reaches the same stage as the first, and its growth ceases in like manner. This accounts for the fact already mentioned (Art. 64), that the younger internodes are shorter than the older.

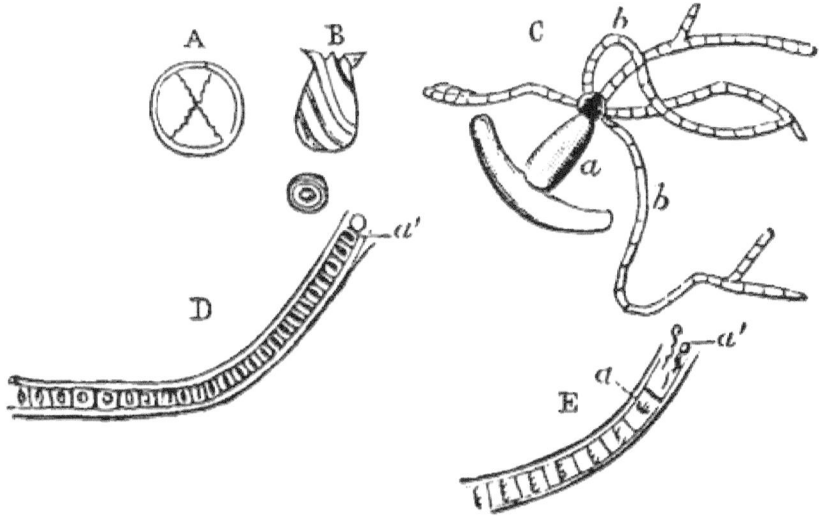

Fig. 32.—A, Antheridium, dividing into eight segments. B, Spore-fruit or nucule. C, Segment of Antheridium, with manubrium a, and filaments b. D, Filament enlarged, showing cellulose sac, segmentation, and the enclosed antherozoids not yet fully developed. E, The same, with antherozoids coiled up in their cells, and some of them, a', escaping.

72. The branch, where it exists, grows from the *axil* of a leaf. Hence if we regard the *leaf* as a *stem*, which indeed it closely resembles in structure, the antheridium will hold the place of a modified leaflet, and the spore fruit that of a modified *branch*, inasmuch as it grows from the axilla of the leaflets (fig. 26). The spore fruit is a bottle-shaped mass, ending above in five little projec-

tions—the termination of the five spiral bands that envelop it. When ripe, the bands leave an opening to the central cell of the spore fruit.

73. Reproduction in Chara.—When the antheridium A (fig. 32.) becomes developed, the segments into which the protoplasm is divided separate from each other, and escape from the cell. C represents such a segment, with the "handle" and *filaments*. The "handle" (or *manubrium*) is the part by which it was united to the other segments when they all formed part of the antheridium. One of the filaments is also represented at D, segmented into cells, each of which bears a spheroidal protoplasm. At a more advanced stage of development the protoplasmic masses are metamorphosed into little spiral filamentous bodies, E, *a*, which ultimately burst from their enclosures, E, *a*, and by means of their cilia find their way by an open passage into the spore fruit B, which they thus fertilise. These ciliated forms are called *antherozoids*. The spore fruit then increases in size, becomes divested of its spiral envelopes, and is perfected into a spore.

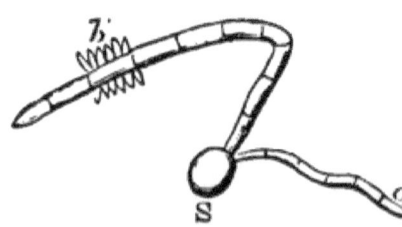

Fig. 33.—Pro-Embryo of Chara. S, Spore germinating; *a, b*, filaments, which give off little projections at *b*, from which the chara takes its origin.

The spore, when sown, first gives off two filaments like hyphæ (fig. 33). One of these, *a*, serves as a temporary root. A cell in the other gives origin to a group of lateral projections seen at *b*, from which at once springs forth the young chara. This temporary structure is called the *pro-embryo*, while that from which the young plant immediately proceeds is designated the *embryo*.

74. Summary.—In treating thus far of the chara, it has been shown that it consists, in the first place, of an axis, with its appendages; and in this structural arrangement we have a foreshadowing of the morphological details of the higher plants. It has also been shown that the chara consists of an aggregation of cells, each having a striking resemblance to protococcus in the more essential details. Each cell of the chara has a sac of cellulose, a protoplasm, or primordial utricle, enclosing a vacuole; chlorophyll floats about in granular masses in the peripheral parts of this utricle. In one point, however, the cell of the chara differs from protococcus. It has a *nucleus*, or, in other words, it is a nucleated cell. Like the mobile protococcus it possesses, in the protoplasm, a power of *contractility*, which may be observed in the movements of the protoplasmic fluid, up one side of the cell and down the other, as well as in the ciliated antherozoids which escape from the antheridia. The cells of chara differ from each other only in minor details—not in any essential characteristic. In the growth of the chara the terminal cell alone undergoes division, first into two, one of which remains single; the other, by lateral fission, gives rise to a family of cells, which again give origin to the lateral appendages. This affords us considerable assistance in the study of the higher plants, as they are all formed from the multiplication and aggregation of *nucleated* cells.

CHAPTER VII.

THE FERN.

75. If we examine an ordinary fern of this country, the common bracken or brake fern, for example, we observe that what at first sight might appear to be a root is not such. It is an underground stem, called also a *rhizome* or *root-stock*. The little filaments that appear hanging from it are the real roots. But many ferns have a stem that rises into the air, such as tree ferns, etc. The fern, then, has an *axis* and *appendages* (fronds and rootlets). In the fern with vertical stem, the fronds are either repeated in "whorls," or they form a tuft at the summit, in the latter case constituting a collection of whorls with suppressed internodes. In the fern with underground stem, these dispositions of the fronds are but seldom observed.

76. In treating of the fern, we must first examine the axis. The outer coat, or *epidermis*, is observed to be of a brownish hue. The general *parenchyma* (or cellular structure) consists of polyhedral cells which are nucleated, and contain chlorophyll and starch granules (figs 34, *b*, and 35, *e*).

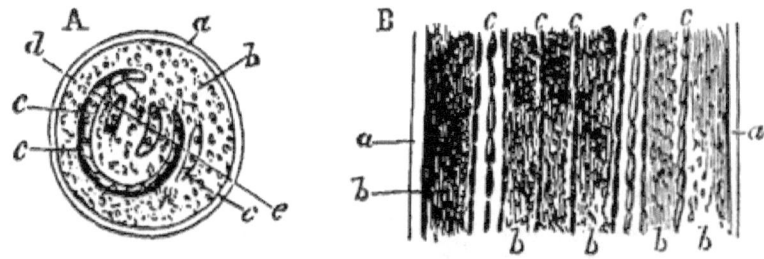

Fig. 34.—A, Transverse Section of a Fern Stem. *a*, Epidermis; *b*, parenchyma; *c*, sclerenchyma, consisting of ducts (annular and scalariform) and fibrous tissue. In each bundle of sclerenchyma, the vessels and ducts occupy the interior, the woody fibres the margin. B, Longitudinal section of same, along the line *d, e*. The same letters indicate the same kinds of tissue.

In K (fig. 35), little projections first show themselves on opposite sides of the cell walls; these, in the progress of growth, soon assume the appearance presented by the *scalariform vessels*, as exhibited in fig. 35, *f*. The *spiral vessels* or "ducts" are produced in a somewhat similar fashion (fig. 35, *g, c*). Outgrowths of granules take place along the inside. These granules, by their coalescence, form a spiral thread, turning round and round inside, like winding stairs in a tower.

77. Tissues.—A *tissue* is a structure so modified as to present a particular appearance, and to serve a particular purpose. All the different tissues in a fern originate in a single growing cell, just as in the chara.

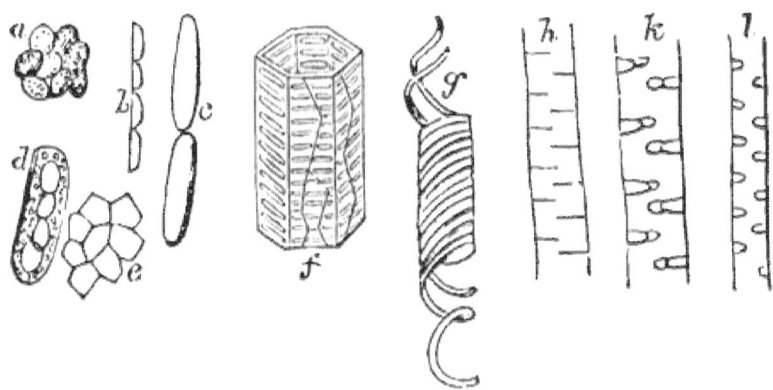

Fig. 35.—HISTOLOGY OF FERN STEM. *a*, Cells of parenchyma; *b*, cells of epidermis; *c*, arrangement of ducts in sclerenchyma; *d*, cross section of fibro-vascular bundle; *e*, arrangement of polyhedral cells; *f*, scalariform duct; *g*, spiral duct; *h*, longitudinal section of scalariform vessel; *k*, mode of growth in same; *l*, mode of growth in spiral vessel.

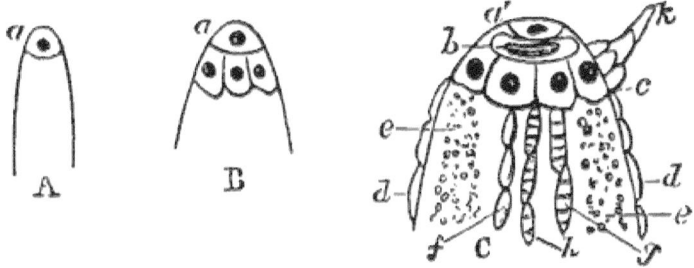

Fig. 36.—DIAGRAM, showing the mode of growth in the Stem of a Fern. A, B, C, Stems of ferns showing successive stages of growth. *a, a, a'*, Terminal cells, the latter just after being produced by division; *b*, a cell which will give rise to an internode; *c*, shows a ring or cluster of cells giving rise to a node; *d*, epidermal cells; *e*, parenchyma; *f*, sclerenchyma; *g*, scalariform vessels; *h*, spiral vessels; *k*, an appendage, originating at the node; *d, e, f, g,* and *h*, all arise from the multiplication and metamorphosis of the "growing" cells.

The diagrams in the annexed figure (fig. 36) show how the one growing terminal cell gives rise, by growth, multiplication, and metamorphosis, to the various tissues to be found in the fern. Hence this plant also is *acrogenous*.

78. The Frond.—In the fern, the leaf bud continues growing in the same manner as the stem does. In this it differs from the chara, and from all flowering plants. In these latter the terminal cell, at a particular stage, is arrested in its further development—becomes barren, so to speak—and the extension of the leaf in that direction is the result of the multiplication and enlargement of those cells which are nearer to the base of the leaf.

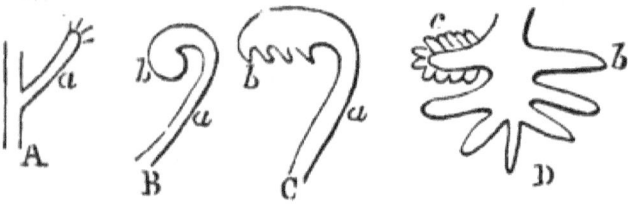

Fig. 37.—A, B, C, D, Stages in the growth of a Fern Frond.

In the fern, on the other hand, the peduncle (fig. 37, A, *a*) grows first. A little projection is then formed B, *b*; this, by the formation of other projections, assumes a toothed appearance C, *b*; by degrees the *lamina* or "blade" is brought forth with its pinnules D, *b*, each of which may again, by the formation of new projections, form secondary pinnules D, *c*, and so on.

79. It has already been observed that carbonic acid is absorbed by the protococcus; that it is decomposed into oxygen and carbon by the chlorophyll granules under the influence of light; and that the carbon becomes "fixed," that is, is retained *as food* to build up the substance of the organism, while the oxygen is evolved. Exactly the same series of operations proceeds in the frond of the fern, and to a less extent in the more deep-seated parts of the plant. But as the individual cells of the parenchyma, which thus act on the carbonic acid, are either at some distance from the free surface of the leaf, or are separated from it by a comparatively thick layer of epidermic cells, some means of establishing a free communication between them and the atmospheric air becomes necessary. This is provided for :—

1. By air openings (*stomata*) in the epidermis of the leaves and stem (figs. 38, 39).
2. By air passages in the interspaces among the loosely aggregated cells of the parenchyma.
3. By *ducts* (*spiral* and *scalariform*), which allow a free communication between those air passages and the more deeply-seated parts of the plant.

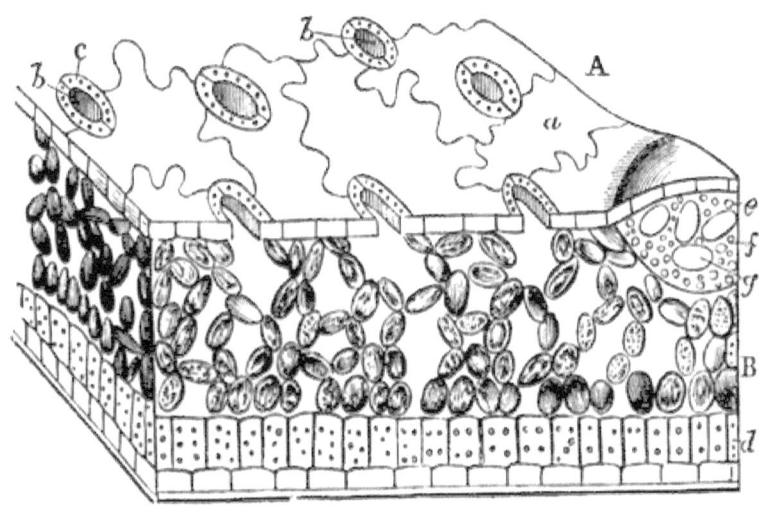

Fig. 38.—UNDER SURFACE AND CROSS SECTION OF AN INVERTED LEAF, to show the functions of stomata. A, Under surface, showing, *a*, cell of epidermis with waving contour; *b*, stoma, having a "guard-cell," *c*, on each side. Three stomata are seen in section, showing how air is admitted into the intercellular spaces; *d*, closely packed cells at upper surface of leaf; *e*, a fibro-vascular bundle cut across showing *f*, woody fibre; *g*, open end of ducts.

80. As it is only in sunlight, direct or diffused, that the chlorophyll can thus effectually de-oxidate the carbonic acid, plants requiring much starch or cellulose in their tissues thrive much more vigorously in bright sunny weather than when it is dark and hazy.* It is said that

* Most ferns luxuriate in shady and damp places. This is because they require for their healthy existence *less* carbonaceous

at night plants absorb oxygen, and exhale carbonic acid, just as animals do at all times. It is to be presumed, of course, that in plants, as well as in animals, decaying tissues requiring, from any cause, to be eliminated must be removed by oxidation, and this both by night and by day. But the quantity of carbonic acid exhaled under sunlight by green plants is very small compared with the amount absorbed by them; as a consequence, the evolution of carbonic acid in the day time is masked by the greater absortion of the same, and hence it is only at night that the former phenomenon is distinctly observable. The leaf, then, is a respiratory organ in the fact that it absorbs oxygen, and gives out carbonic acid; while its office in absorbing carbonic acid, decomposing it, and "fixing" the carbon, constitutes it an *alimentary* or *assimilating* organ as well.

81. The same mechanism of *stomata, air passages,* and *ducts,* for the admission of atmospheric air, assists the plant in the absorption and exhalation of moisture. The power of absorption depends upon the thinness of the epidermis, and on the number of stomata with which it is provided. Hence, for both reasons, the under surface of the leaf will absorb much more than the upper. It is only when the soil is so dry that the roots cannot properly discharge their functions that leaves absorb moisture, either as aqueous vapour by the stomata, or as fluid matter by the hairs and cuticle. The moisture so absorbed is conveyed to the roots by the *ducts,* which always keep up a free communication between the leaves and all other parts of the plant. In ordinary cases, however, the crude sap, when it enters the leaves, after ascending through the root and stem, becomes at once exposed to the action of the air contained in the intercellular spaces, and which communicates with the external air, through the stomata. The consequence is that its superfluous moisture—often a very large part—passes into the atmatter, and *more* water, than is needed for those plants that lay up much starch, sugar, or woody matter, within their substance.

mosphere, and the sap becomes *inspissated*. This difference in the densities of the sap in the leaves, and of that in the stem and roots, causes the *ascent* of the crude sap by the physical operation technically termed *endosmose*. The difference being constantly maintained, under ordinary circumstances, by exhalation from the surfaces of the leaves, the sap must in consequence constantly ascend. The quantity of moisture exhaled must vary, of course, with the hygrometric state of the air, being greater while the air is dry, and less when it is moist; but the stomata have a kind of self-regulating apparatus (fig. 39, *a*) which renders the comparative densities of the sap in different parts of the plant tolerably uniform.

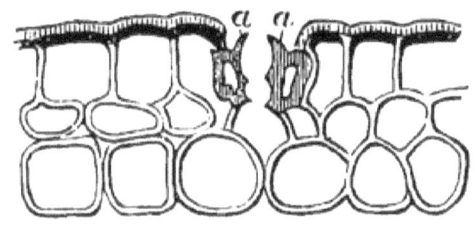

Fig. 39.—SECTION OF STOMA, showing "lips," *a*, which open or close, so as to regulate evaporation from leaf.

Other important functions of the leaf will be referred to when we come to treat of "flowering" plants.

82. **Reproduction in Fern.**—There appear on the under side of some of the pinnules, rounded brown spots, or aggregations of little brown bodies, which have a very singular appearance. Each patch is a *sorus* (sometimes covered in by a membrane called an *indusium*), and the little brown bodies of which it is constituted are called *sporangia*, or *spore cases* (fig. 40, A, B). The sporangia are developed from the epidermal cells. An elastic ring (*annulus*) surrounds each sporangium (fig. 40, B, *a*) and assists in its dehiscence. The cells which escape become divided into four segments F, *b*, each of which takes two coats, the outer, called *exosperm;* the inner, *endosperm.* The segments are set free immediately upon the bursting of the sporangium. The production and perfecting of these spermic bodies is the ultimate act of the life of the frond, after performing which, it perishes. In some

52 GENERAL BIOLOGY.

Fig. 40.—A, Portion of Frond of the common Shield Fern (*Aspidium filix mas*), to show sori, *a*, covered by an indusium. B, Sporangium yet unbroken; *a*, annulus; *b*, spore case. C, Section of B, along the line *c*, *d*; *a*, annulus; *b*, cavity containing spores. D, a sporangium after bursting, with one spore yet remaining in the capsule. E, annulus straightened by its own elasticity; a spore is seen at either end. F, Enlarged view of ruptured sporangium; three spore cells are seen, each containing four sporules. G, Royal Osmund fern; *a*, vegetative frond; *b*, reproductive frond, bearing spores.

ferns, such as the common Hard Fern, and Osmunda Regalis, special fronds are set apart for the production of spores (fig. 40, G, *b*).

83. If one of these spores be sown on a glass plate, and kept moist with water in which it can grow, and kept covered by a glass case, the exosperm of the sporule bursts, and a long filamentous body, like the pro-embryo of the chara, is produced (fig. 41, *c*). An expansion *d* is then formed which gives out little rootlets *e*, somewhat resembling those of a *lichen* or *marchantia*. This expansion is called a *prothallium*. It is of a brownish hue, and its cells contain chlorophyll. On its under side are produced two kinds of bodies, one called *archegonia*, the other *antheridia*.

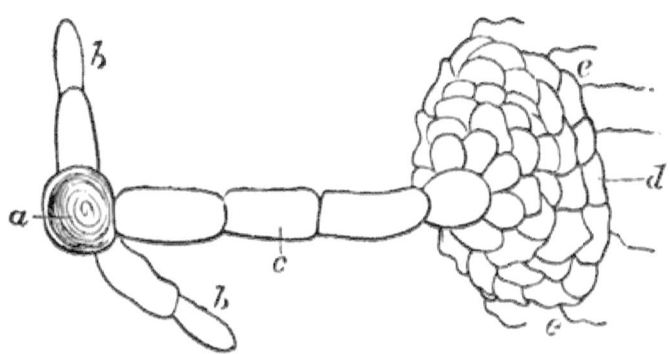

Fig. 41.— PROTHALLIUM OF FERN. *a*, Spore; *b*, *b*, first roots; *c*, prolongation from spore (like pro-embryo of chara); *d*, prothallium; *e*, rootlets.

The archegonium, when ripe, opens as at B (fig. 42). The protoplasm of the antheridium breaks up into minute cells, from which ciliated antherozoids *b*, *c* are produced. These antherozoids pass into the archegonium through the opening, and fecundate the cell *g*. The latter then rapidly increases in size, divides itself, and becomes a true embryo, *i.e.*, develops new cells, some of which go to form a stem, others a frond, and others the rootlets. In other words, a true embryo always gives rise to a *bud*.

84. In tracing the process of multiplication in ferns,

we find (1) that the frond produces spores; (2) that these spores, when sown, give rise, *not to a fern,* but to a *prothallium:* this is *asexual generation;* (3) the prothallium produces archegonia and antheridia by which *sexual* fecundation takes place; and (4) the embryos thus produced do not bring forth a *prothallium,* but a *true fern.* Thus we have two forms of the plant, one succeeding the other, but wholly independent of each other so far as structure, etc., is concerned. We have—

1. An *embryo,* produced *sexually,* bringing forth a *true fern.*

2. A *spore,* produced asexually, bringing forth a *prothallium,* different in many respects from the fern. This is an excellent example of what is known in biology as alternation of generations.

Fig. 42. Fig. 43.

Fig. 42.—A, Archegonium closed. B, Archegonium opened, admitting antherozoids, *c; g,* germ cell. C, Antheridium containing cellules, *a,* which, after their escape from the sac, liberate ciliated antherozoids, *b, c;* these finding their way by means of their cilia to the germ cell *g,* in B, fertilise it.

Fig. 43.—EMBRYO FERN. *a,* Archegonium; *b,* frond; *c,* stem or rhizome.

CHAPTER VIII.

MORPHOLOGY OF A FLOWERING PLANT—THE BEAN.

85. The common bean is a good example of an exogenous plant. Like the chara it possesses an axis with its appendages. The axis may be divided into what is

under and what is over the ground, the former portion being called the **root**, the latter the **stem**. The appendages of the root are the rootlets; of the stem, branches and leaves. The rootlets are irregularly disposed, but the arrangement of the leaves presents a great degree of regularity. In the bean they are disposed in the fashion which botanists term alternate. In most plants the stem is round or cylindrical, but in the bean it is angular or square. As in chara, the internodes become smaller as the stem is younger. As we approach the summit, we find them to be very small; while at the summit itself, we observe that the appendages are crowded upon one another in consequence of the non-development of the internodes. The terminal bud is also in several respects similar to that in chara. From certain parts branches are given off in the axils of the leaves. The branch, when present, is simply a repetition of the axis, speaking generally; but the flower, which is in its nature a branch differs a good deal from the stem.

86. In speaking of the parts of a **flower**, we call that portion which is nearest the stem the *dorsal* part, that which is opposite to this is the *ventral* part. At the head of the flower-stalk is a little cup called a *calyx*, formed of green leaves united together, and terminated in five points at its free extremity (fig. 44, A). In the flower of the bean, the calyx is formed of *five* pieces fastened together, each piece being called a *sepal*. The second whorl, the *corolla*, is also made up of five parts, called *petals* (fig. 44, B). The upper one, which is the largest, and folds over the second pair, is called the *vexillum* B, C, *a;* the lateral pieces are called *alæ* B, C, *b;* and the under two, which are fastened together at their inferior edges, are called the *carina* or the keel B, C, *c*. (See also D, E, F.) Inside the whorl, constituting the corolla, we notice a hollow sheath, bearing at its outer end ten short slender filaments, each of which supports a sort of double bag, containing minute grains, or dust-like bodies, called *pollen grains* G, *a*.

Enclosed again in this sac-like sheath is a hollow body called the *carpel*, enclosing, in its turn, minute cell-like bodies called *ovules* H, d.

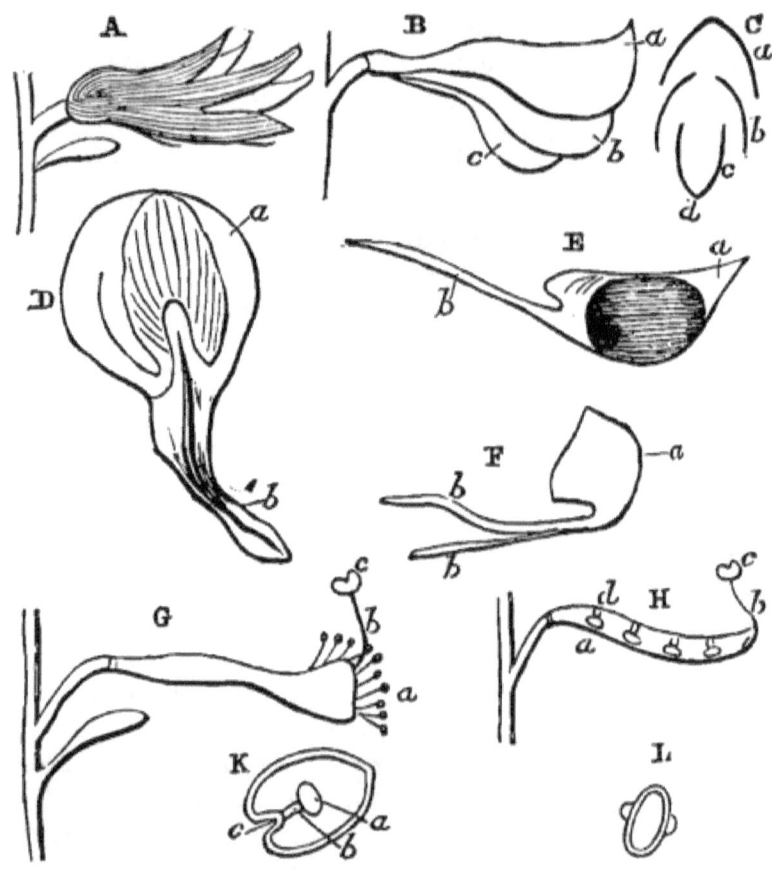

Fig. 44. PARTS OF A BEAN FLOWER. A, Calyx, 5-toothed. B, Corolla, showing a, vexillum; b, ala; c, carina. C, Cross section of corolla; the carina is formed of two petals united in their limbs at d. D, E, F, Vexillum, ala, carina, respectively; a, limb; b, unguis. G, Flower, after calyx and corolla have been removed; styles of stamina united into a membrane sheathing the carpel; a, anthers; b, style; c, stigma. H, Stamina removed to show carpel a, containing ovules, d; b, style; c, stigma. K, Transverse section of carpel; a, ovule; b, funiculus; c, placenta. L, A pollen grain magnified.

We have thus:—

$$10 \begin{cases} 5 \text{ sepals} & = 1\text{st whorl.} \\ 5 \text{ petals} & = 2\text{nd ,,} \\ 5 \text{ stamens} & = 3\text{rd ,,} \\ 5 \text{ stamens} & = 4\text{th ,,} \\ 1 \text{ pistil}* & = \text{axis.} \end{cases}$$

We thus observe that the number five regulates, in a large measure, the arrangement and distribution of the parts of a flower. We may regard each system as a whorl with internodes undeveloped. Indeed, all the parts of a flower are homologous to the ordinary appendages of the stem.

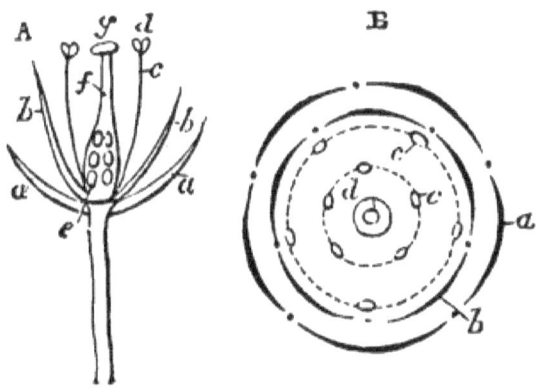

Fig. 45.—A, Sectional view of the flower, showing the vertical disposition of the whorls; a, sepal of calyx; b, petal of corolla; c, filament of stamen; d, anther of stamen; e, ovary of pistil; f, style of pistil; g, stigma of pistil. B, Plan of the typical flower of an exogenous plant, showing the horizontal disposition of its parts; a, sepal; b, petal; c, c, stamens in two distinct whorls; d, carpel or ovary, enclosing an ovule, attached by its funiculus.

87. The Stem—Stomata.—The angular lines in the stem of a bean plant have a spiral turning similar to what may be observed in chara. The surface appears green, owing to the presence of chlorophyll, but the

---

* The *pistil* of a flower may consist of *one* carpel (as in the bean), or of several carpels united by "coalescence." In the latter case, the pistil is said to be *syncarpous*.

outer layer, or epidermis, is in reality transparent. If a portion of the outer integument be scratched off and examined, it will present the annexed appearance A, (fig. 46); while, if examined edgewise, the outline at B will be observed, showing that the cells are flattened. On looking at the *flat* side of the piece, figures like that marked C may be noticed, formed of two kidney-shaped cells, touching each other, and allowing a hollow to exist between them. This space is called a *stoma* (plural, *stomata*). Under the microscope, the central space seems quite black, owing to its containing air, which, by filling a hollow cavity in the water, causes the latter to act on the principle of "total reflection," in consequence of which it can transmit no light. A similar reason may be assigned for the black appearance of air bubbles under the microscope. D represents a section through a stoma, showing how it opens to the surface. It is by means of stomata that air is admitted to all parts of the plant.

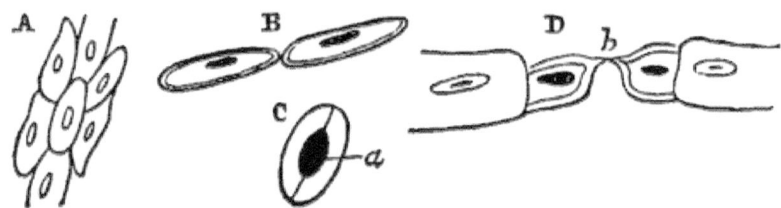

Fig. 46.—A, Epidermal cells; B, the same, seen edgewise; C, two kidney-shaped cells enclosing a stoma, *a;* D, section showing how stoma opens and admits air at *b*.

There always exist intercellular spaces which allow access of air in all directions, for the benefit of the subjacent tissues. Spiral vessels and dotted ducts also contain air. There is no cell in the whole structure of a plant quite shut out from the action of the atmosphere. Even in the very interior of the stem of the bean there is a means provided for the admission and circulation of air, the stem being in fact hollowed out for that purpose. In its earlier stages, the bean stem is quite solid, and is entirely composed of cellular tissue. As growth proceeds,

THE STEM—STOMATA. 59

bundles of woody fibres are developed in the midst of the parenchyma. The growth of the parenchyma in the interior does not keep pace with the peripheral growth of the stem; the newly developed woody bundles causing the latter to be distended considerably. The central mass

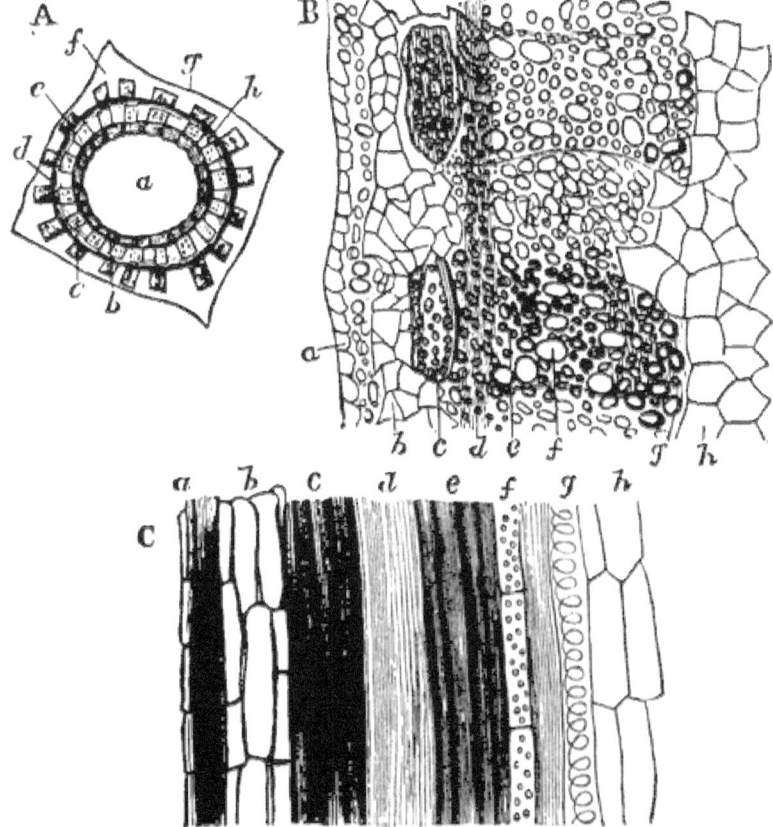

Fig. 47.—A, Cross section of bean stem, showing the arrangement of the tissues; *a*, hollow interior; *b*, medulla or pith; *c*, inner fibro-vascular bundles; *d*, cambium layer; *e*, outer woody bundles (of cortex); *f*, parenchyma of cortex; *g*, epidermis; *h*, medullary ray. B, C, Transverse and longitudinal sections of stem, magnified; *a*, epidermis; *b*, parenchyma of cortex; *c*, woody fibre of cortex; *d*, cambium layer; *e*, woody fibres of stem; *f*, dotted vessels; *g*, spiral vessel; *h*, pith; *k*, medullary ray (B).

of parenchyma is hence pulled outward in opposite directions; the cells are torn asunder or ruptured, and a hollow is formed. As growth proceeds, the same influences continue to act, the hollow becomes wider and wider, many of the loose parenchymic cells disappearing by becoming absorbed into the body of the stem.

88. Commencing at the pith, the outward order of the tissues is (fig. 47, A, B, C):—(1) pith; (2) spiral vessels; (3) dotted vessels and woody fibre; (4) cambium layer; (5) woody fibres (liber) and vessels of cortex; (6) parenchyma of cortex; (7) epidermis. The woody fibres and vascular bundles are arranged round the pith in wedge-shaped masses, with spaces between filled with parenchyma like the pith. These last are the *medullary rays*.

CHAPTER IX.

PHYSIOLOGY OF A FLOWERING PLANT—THE BEAN.

89. All this variety of tissue is a modification of ordinary cells. The growth, in the first instance, takes place by a terminal bud as in chara. But there is this difference, that whereas in chara the number of cells below a terminal bud is never increased, the case is different in the bean; for, concurrent with the multiplication of the cells of the terminal bud, the cells of the cambium layer produced by it become also multiplied, and increased in growth; so that the tissues below, already formed, or in course of formation, have new matter added to them through the vital action of the cambium cells, which are always renewed from above downwards. In fact, in the bean, or indeed in any flowering plant, there is no one *cell* which may be called " terminal," the terminal bud being always made up of a *number of cells* in process of growth and multiplica-

tion. It is plain, then, that the stem increases in *width* as the plant grows, from the additions made to it by the cambium layer. As these additions are made on the *outside* of the main part of the stem, the plant is called **exogenous**, *i.e.*, *growing on the outside*. But the bark increases *from the inside*.

90. The general appearance of the axis of an exogenous plant is a double cone — one cone representing the stem, the other the root; the growing part in either being bathed in the cambium fluid.

Speaking generally, it may be said that the **root**, or descending axis, is constructed on the same principle as the stem. It is, like it, covered by a layer of epidermis; but, unlike it, this epidermis possesses no stomata. It is important to mark this distinction between the root and the stem. The central parenchyma answers to the pith. The roots also possess partitions of parenchyma which are analogous to the medullary rays of the stem, and also another part corresponding to the bark—the *epiblema*. The root, like the stem, has spiral vessels, dotted ducts, pitted cells, and cambium.

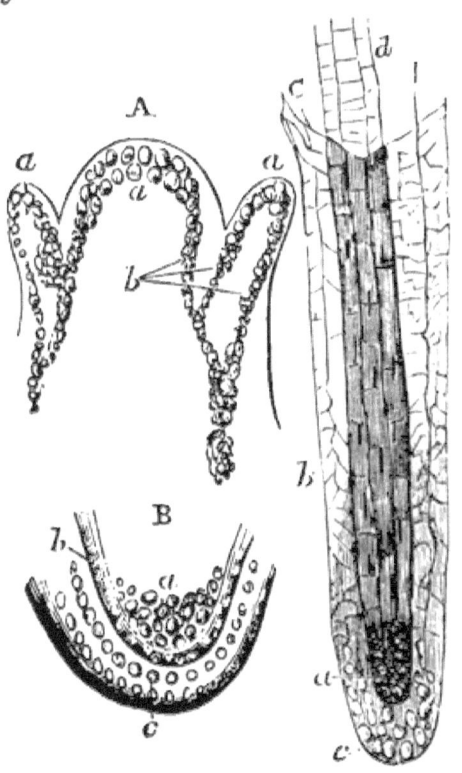

Fig. 48.—A, Mode of growth in stem B, In root. A, *a*, Growing cells in stem, which multiply by fission; *b*, cambium, elaborated by growing cells. B, *a*, Growing cells in root; *b*, cells produced by growing cells; *c*, cap (pileorhiza). C, Root of duckweed (magnified); *a*, growing point; *b*, root-sheath; *c*, cap; *d*, root.

The woody fibres, however, are differently disposed in the root. The appendages also of the root differ from those in the stem. There are *no leaves*, but *rootlets* instead, and these are irregularly arranged along the root which tapers to the extremity.

91. The mode of growth in the root differs very materially from that observed in the stem. It is difficult to observe this in full grown roots, but in embryo roots it is very easily noticed, inasmuch as the terminal joint is available for the purpose of examination.

In the growth of the stem, the terminal cells a (fig. 48, A) multiply and enlarge, while they give new cells to the cambium layer, by which the growing portion beneath is increased in size. In the root, on the other hand, the multiplying cells are *not quite* at the extremity. The original growing cells, a (fig. 48, B, C), of the radicle make a kind of cap for themselves, which cap grows by additions made to its interior. These *push out* the layers external to them much in the same way as the cuticle of the skin grows by the successive formation of epidermic cells, of which the oldest are on the outside. In this way the cells at a go on multiplying and increasing, pushing the cap c before them, which thus affords efficient protection to the newly-formed tissue. The *lengthening out* of the root is thus effected entirely by the group of growing cells at c.

92. The appendages to the stem are the **leaves** (inclusive of *bracts* and *stipules*) and **branches,** to which latter class the flowers may also be said to belong.

If we examine the leaf of a bean, we find it to be made up of a broad, thin mass of parenchymic tissue, enclosed within two layers of epidermis, upper and under, with "veins" of fibro-vascular tissue branching out through it, and communicating with the fibro-vascular bundles of the stem through the petiole or leaf-stalk. The vein which forms the continuation of the petiole, and which runs down through the middle of the leaf, is called the *midrib.* In strictness it is to the branches and network proceeding from this mid-rib that the term "veins" is applied. The epidermis of the leaf is like that of the stem, but the cells have a more waving contour. The epidermis and parenchyma of the leaf are continuous with those of the stem. In fact, the leaf is simply an outgrowth of the principal structures of the stem itself. It differs from the stem partly in the arrangement of its structures, and partly in its mode of growth.

93. In the stem the growing terminal cells retain their full activity, continuing to divide during the life of the plant as in the stem of chara; but in the leaf, as in the leaf of chara, the terminal cell *when once formed multiplies no longer*. In ferns the reverse of this takes place in the growth of the frond, the apex of the leaf being the part most lately formed. But in the higher plants the growth resembles that of chara, the activity of the terminal bud of the leaf being arrested, and the tissues formed round it being *pushed out* by the growth of the part nearest the stem. In the fern, therefore, the apex of the frond is its youngest part, and the peduncle the oldest; whereas, in the higher green plants, the point of the leaf is its oldest part, the leaf-stalk being the youngest. At one time it was thought that this latter mode of growth in the leaves was general in all plants; but we have seen that such is not the case with the fern.

94. If a transverse section of a leaf be made, it will be found to exhibit somewhat the appearance presented in the accompanying diagram (fig. 49). The upper layer is formed of epidermic cells, provided at intervals with stomata. Beneath these there are one or two layers of cells *vertically* arranged. Under these, again, the parenchyma becomes quite loose, being formed of cells arranged in an irregular stellate order, with large intercellular spaces between; and on the under side of the leaf there is another layer of epidermic cells more largely provided with stomata. There is also a section of a *vein* represented, showing the openings of the spiral vessels, dotted ducts, and woody fibre. Inasmuch as the interspaces

contain air in large amount, their appearance under the microscope is dark; or else they give rise to those black spherical masses well known as air bubbles. The vertical and stellate cells are filled with granules of chlorophyll, to which the green appearance of the leaf is due.

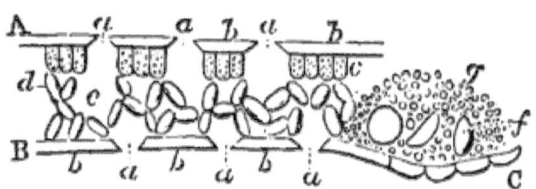

Fig. 49.—DIAGRAMMATIC SECTION OF A LEAF. A, Upper surface; B, under surface; a, a, stomata, opening between b, b, epidermic cells; c, vertical cylindric cells; d, cells of parenchyma loosely arranged; e, intercellular space or air passage. C, Transverse section of "vein," or fibro-vascular bundle, showing f, spiral duct, and g, woody fibre.

95. Branches are developed from *buds* arising from the axils of the leaves. A cell a, arising from the undifferentiated tissue always to be found near the axil, projects; soon other little projections are developed from it laterally, as seen in the figure, and thus the bud is formed. When the flow of the cambium layer becomes active, the bud develops itself into a branch, which is merely a repetition of the axis, and grows like it. There is always undifferentiated tissue near the axils of leaves: the reason is not known, but such is the fact.

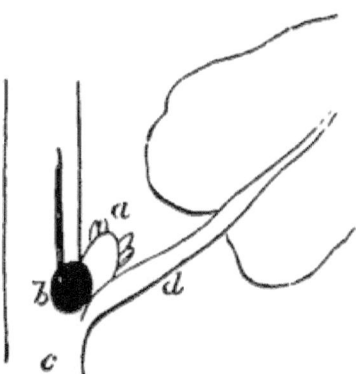

Fig. 50.—DEVELOPMENT OF BUD. a, Bud sprouting from *axilla* of leaf; b, cambium; c, stem; d, leaf-stalk.

96. The Flower of the bean is a modified branch with its appendages. The latter are seen to consist of a series of whorls, arranged in the following order, commencing at the outside:—

1. 5 sepals united at their edges.
2. 5 petals, two of them united, three free.
3. 10 stamens, with their filaments united in a tube.
4. 1 carpel, with its style and stigma.

Inside the carpel are contained the ovules.

The full-grown bean is the matured carpel. Sepals and petals are repetitions of the leaves, the main difference between them consisting in the nature of their colouring matter.

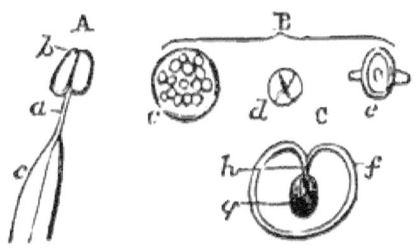

Fig. 51.—A, Stamen of bean; *a*, filament; *b*, anthers filled with pollen. The filaments of all the stamens coalesce to form the "staminal" tube *c*, which envelops the carpel. B, *c*, Cross section of anther, showing "pollen cells" in the interior; *d*, one of such cells detached, and breaking by cleavage into four segments; each of these segments invests itself in one or two coats of cellulose. The parent cell finally becomes ruptured, setting free the four cellules, each of which becomes a pollen grain, of which *e* is a cross section. C, Transverse section of carpel, showing the carpellary leaf *f*, and one ovule *g*, attached to the placenta *h*.

97. The Stamens, which are also modified leaves, are the male apparatus of the flower; they consist of two little bags *b* (fig. 51, A), called anthers, supported on a stalk *a*. These anthers contain cells, which divide themselves into four parts, each of which surrounds itself with two coats, called respectively *intine* and *extine*, and thus becomes developed into a *pollen grain*. These pollen grains, when mature, after rupturing the anthers containing them, fall on the stigma, and are there detained by the hairs with which it is covered, and the viscid matter secreted by it (fig. 52, A). The intine then protrudes through apertures in the extine (fig. 52, B, *d*), forming long tube-like projections, which thread their way through the tissues of the style and placenta till they ultimately reach the ovules, which they fertilise. Before describing how this is effected, it will be necessary to explain the structure of the carpel and its contained ovules.

98. The Carpel (fig. 51, C) is homologous to a leaf

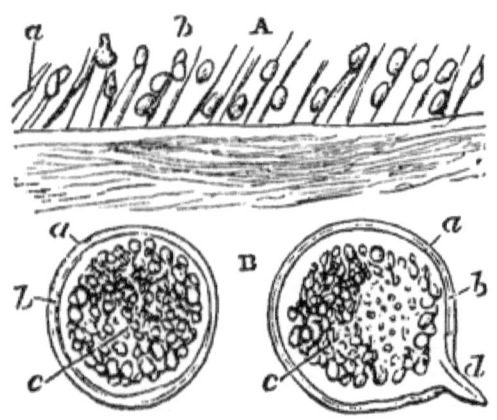

Fig. 52.—A, Brushes on stigma of sweet pea, showing *a*, hairs; *b*, pollen grains attached; B, pollen grains magnified to show *a*, extine; *b*, intine; *c*, granular protoplasm; *d*, commencement of pollen tube where the intine is seen protruding.

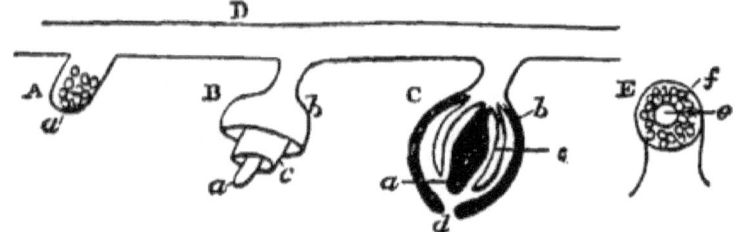

Fig. 53.—A, B, C, Stages in the growth of an Ovule. *a*, Nucleus; *b*, primine or outer coat; *c*, secundine, or inner coat; *d*, foramen, or micropyle. Fig. C is shown in section; D represents the placenta; E, nucleus, showing *e*, embryo sac; *f*, cells of endosperm.

curving inwards till its two margins become attached. Along this line of attachment, called the *placenta*, the ovules are arranged. The first appearance of the ovule, which is a true bud, is as a cellular projection (called the *nucleus*) from the placenta (fig. 53, A). After a little time, two other projections appear on the surface of this, B, *b, c*, and by their growth soon envelop the nucleus. As the growth proceeds, the outer envelope tends completely to cover the nucleus, but a small space is left uncovered at *d*, called the foramen or micropyle. The growth of *c* also goes on uniformly; but it, too, leaves the micropyle uncovered. The nucleus inside of these coats has an appearance somewhat like E. It contains the *embryo sac e*, surrounded by a mass of smaller cells *f*, called the *endosperm*.

99. After the pollen grain falls on the stigma it begins to throw out a projection upon one side (fig. 52, *d*). This grows longer and longer, down through the interior of the style, till at last it reaches an ovule. It passes in through the micropyle and comes into contact with the wall of the embryo sac (fig. 54, A). The protoplasm of the pollen grain is then absorbed into the *germinal cell* or *vesicle* contained within the sac. The vesicle thus becomes fecundated, after which it rapidly enlarges by the absorption of the cells around it. It throws out a projection, which, becoming lengthened by the formation of new cells, at last gives

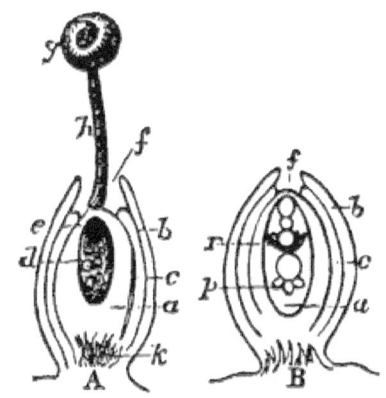

Fig. 54.—A, Diagram, showing the impregnation of the ovule. *a*, nucleus; *b*, primine; *c*, secundine ; *d*, embryo sac ; *e*, germinal cell; *f*, foramen, or micropyle; *g*, pollen grain; *h*, pollen tube. The style through which the pollen tube finds its way is not represented. B, Development of germinal cell after impregnation. Letters same as in A, except *p*, plumule ; *r*, radicle. Surrounding the embryo is the endosperm.

of a radicle and a plumule. In this way the embryo is formed, and the *seed* becomes matured (fig. 54, B, *r, p*).

The essence of impregnation consists in the fusion of the protoplasm of one kind of cell with that of another kind. This actual fusion has not yet been observed in the embryogeny of phanerogamous (*i.e.*, flower-bearing) plants; yet, as the production of an embryo—always the result of impregnation—is notably the effect of the action of the pollen on the pistil, we must needs infer that such fusion actually takes place.

In the development of the embryo of the bean, the endosperm is wholly absorbed into its substance—principally into that of the *cotyledons*, which thus hold a temporary supply of nutriment for the young plant after it begins to germinate. In many other seeds, however, the endosperm is not thus wholly absorbed, but becomes elaborated into a substance termed *albumen** (consisting largely of starch and gluten), which affords a temporary nourishment for the young plant till its radicles can draw from the soil the materials proper to its subsistence. Seeds thus provided with an endosperm are hence called *albuminous*, while those wanting it are termed *ex-albuminous*.

100. An ordinary bean seed may easily be separated into two parts, each of which is a *cotyledon* (fig. 55, B, *c*). At their junction is the embryo with its *radicle* and *plumule b, a*. The radicle always grows outward towards the micropyle, the plumule in the opposite direction. The cotyledons serve two purposes: they perform the office of temporary leaves, and they supply the young plant with nourishment (fig. 55, C, *c*).

101. It is thus seen that the fusion of the two protoplasms—those of the pollen grain and embryo cell, results in a *seed*, an organism which differs from a spore in this, that it is not only capable of producing a young plant, but actually contains it ready formed, though but yet in an immature condition. This mode of propagation in plants

---

* This must not be confounded with the chemical substance bearing the same name.

is called *gamogenesis* or *sexual reproduction*—the stamens and pistils being respectively regarded as the male and female organs by which the seed, with its contained embryo, is produced.

There is, however, a second mode by which the multiplication of most flowering plants may be effected, more especially of those whose growth continues for more than one season. It has been already stated (Art. 95) that a leaf-bud will give origin to a branch, which is merely a repetition of the stem, and produced from it as an outgrowth in a manner similar to the *gemmation* (Art. 12) of the torula. A branch, then, may be regarded as a young stem, developed from a bud, and sending down fibres *within the body of the parent* till they reach the soil, and become roots, which then draw up nourishment for both parent and offspring, but principally for the latter. According to this theory, a tree, shrub, or other branching plant, may be regarded rather as a colony of plants (analogous to coral polypes), all living in common, but each drawing nourishment from the soil and air, and

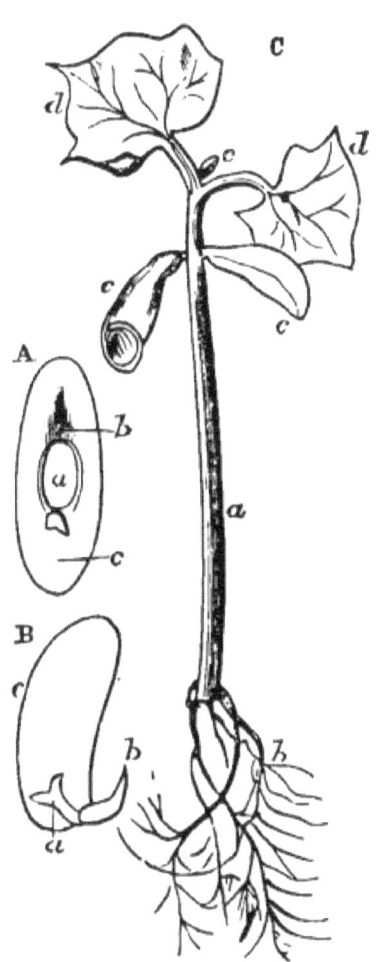

Fig. 55.—A, Seed of bean; *a*, *hilum*; *b*, micropyle; *c*, cotyledon. B, Cotyledon, *c*, of bean, the test being removed; *a*, plumule of embryo; *b*, radicle. C, Young bean plant; *a*, stem; *b*, radicle; *c*, cotyledon; *e*, terminal or growing bud.

digesting it for the common benefit of the family group.

But the bud, and the young plant which springs from it, may also become separated from its parent stem—just as young torulæ do—and live independently. In nature, the buds of strawberry runners, potato tubers, hyacinth bulbs, etc., become detached in this way and give rise to independent plants. The same result may be brought about artificially in various ways; as, for instance, by bending a stem or branch so as to bring a bud in contact with the soil; or by cutting off a portion of the plant containing a bud, and planting it in the earth, or in the cambium layer of another plant of the same, or closely allied, species. In all these cases the plant is propagated *asexually*, and this mode of reproduction is hence called *agamogenesis*.

102. Ferns do not bear flowers, the propagation of their kind being brought about, in the first instance, *asexually* by spores developed in the epidermis of the frond; and then *sexually* by the union of the two protoplasms produced in the antheridia and archegonia—the male and female organs of the prothallium. If we institute an analogy, then, in this matter, we find that the *bud* of the flowering plant corresponds with the *spore* of the fern, both propagating asexually. As the prothallium is derivêd from a spore, and develops antheridia and archegonia, so also the *flower* is derived from a *bud* (being, in fact, a modified branch), and develops pollen grains and ovules. The union of the two former gives true embryonic ferns; that of the latter produces seeds, each containing an embryo (flowering) plant.

The modes of fertilization are, however, slightly different. In the prothallium of the fern, the antheridium *opens*, but remains fixed, and the protoplasmic matter comes out in the form of ciliated antherozoids, which, by their power of independent motion, reach the germ cell in the archegonium. The pollen grain, however, becomes itself detached, and falls upon the stigma. But instead of opening to liberate the protoplasm (here called fovilla), it sends out a long tube which works its way among the tissues of the style, till at length it enters the ovule by the micropyle, and reaches the embryo cell of the nucleus.

CHAPTER X.

NUTRITION IN PLANTS.

103. THE annexed diagram (fig. 56) of a plant structure shows the general arrangement of the different tissues. Air passages are both intercellular and vascular. The roots are supplied with water containing carbonic acid, air, and oxygen, in addition to the mineral and nitrogenous matters contained in the soil. The parts that are green, or contain chlorophyll, are made up of cells that live like protococcus, while the parts removed from the action of light have a vitality more resembling torula. The great mass of the chlorophyll resides in the cells immediately under the epidermis. The inner parts, forming the root and

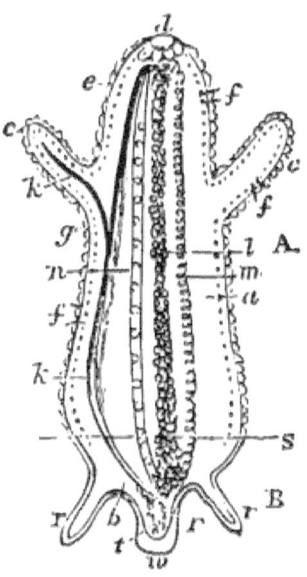

Fig. 56.

Fig. 56.—DIAGRAMMATIC SECTION OF A FLOWERING PLANT, showing the different tissues. A, Ascending axis; B, descending axis. s, Surface of soil; c, c, appendages; d, growing point of stem; e, epidermis; f, f, stomata; g, layer containing chlorophyll (marked by the dotted lines); k, k, woody fibre; l, m, n, pith, spiral vessel, and dotted duct—all air passages; r, r, roots; t, growing point of roots; w, cap (pileorhiza).

central stem, contain no chlorophyll, but are *invested* with a layer of chlorophyll structures, which gives rise to the green colour of plants.

104. The soil is formed of earthy substances, as sand, clay, lime, with *humus* or decomposing organic matter. At one time it was supposed that the latter ingredient was essential to the life, at least, of the higher plants. But it has been frequently observed that plants may grow healthfully in a soil which contains no organic matter whatsoever. Plants growing on bare walls, such as the common house-leek, and the cactus, which grows in the clefts of the rock, or in arid sands, prove that the presence of organic matters in the soil is not necessary to vegetable existence, if the proper plasma be supplied by inorganic materials. This fact may also be proved, experimentally, by placing a germinating bean in a solution containing, in 1000 parts of water, about two parts of the following ingredients:—

Salts, containing—
Sulphur,   Iron, with
Phosphorus,   Nitrogen,
Potassium,   Hydrogen,
Calcium,   Oxygen.
Magnesium

These elements may be presented in the form of—

Potassium Nitrate,
Iron Phosphate,
Magnesium Sulphate,
Calcium Phosphate,
Water.

If exposed to sunlight and air, the bean thus treated will grow just as well as if it were planted in the soil. There is no carbon amongst these ingredients; but the plant, nevertheless, will contain carbon in abundance. The only source is the carbonic acid of the atmosphere; but this is sufficient in quantity to supply carbon to all the vegetation in the world.

105. The only structure capable of effecting the chemi-

cal changes necessary, is the chlorophyll. Hence it is essential that the materials which supply food to the plant should be carried up to the leaves, where the chlorophyll acts upon them. It is not precisely made out in what structure of the plant the carbon is deposited immediately after being abstracted from the carbonic acid; but it is supposed to be first stored up in the starch granules.

106. In the conveying of the sap to the leaves, there are two distinct forces employed:—

    1. A *pulling* force.
    2. A *pushing* force.

The evaporation which constantly takes place at the surface of the leaves is sufficient to account for the former. Cells relatively dry draw the fluid from adjoining cells which are more moist; these again abstract it from their neighbours, still moister; and as the cells in the leaves are constantly deprived of part of their moisture by evaporation, they react on the cells nearest them, and they, again, on cells more moist; and this process of "handing up and taking from below" goes on till the cells of the spongioles are, at length, deprived of part of their moisture, and compensate themselves for the loss by drawing more moisture from the soil around them.

107. The pushing force (*vis a tergo*) differs greatly in different plants. In some it is very considerable. The "bleeding of the vines" in early spring is a good example. The *pushing* force is often equal to the pressure of one atmosphere. It resides in the spongioles of the roots. When we examine it closely, we find that in the long run it depends on the principle of endosmosis.

If a tube, covered at the lower end by an animal membrane, and containing a solution of *gum*, be immersed in a vessel containing water, we find that there is a tendency in the two fluids to intermingle, so as to become homogeneous, or of the same density. There is, consequently, a crossing of both fluids through the membrane.

But as the latter is more readily permeable by the water than by the gum solution, more water will pass into the tube than there will of gum outward. If the tube contained a perfect *colloid*, there would be a great indraught of water, while there would scarcely be any outdraught of colloid. Hence the water after passing into the tube would rise through it with very great force. All this would go on until either both fluids became of the same density, or until the colloid became fully saturated. There is thus a limit to the action of the force created by endosmosis. But we find that the pushing force in the plant is continual. The real explanation is, that as the plant grows a new series of cells in the spongioles come into action, and the operation is repeated.

108. The course of the sap through the plant has been determined by using a logwood dye, and causing it to ascend along with the sap; the tissues thus become coloured, and the course of the sap may be traced. It first passes upward through the vascular bundles, and then diffuses itself through the intermediate spaces, and subsequently through all the growing parts, more especially in the leaves.

109. It is only the *growing* cells that can *manufacture* protein—a substance essentially necessary to the existence of the protoplasm in all cells. The sap, then, must, after being elaborated in the younger or *growing* cells, pass downwards to nourish the older cells, or such as are simply adding to their substance. Thus the portion of protein over and above what is required for the development and growth of the young cells, is distributed among the others, and in a form very much resembling the protein of animal tissues.

CHAPTER XI.

EXOGENS AND ENDOGENS—MODIFICATIONS OF LEAF.

110. THUS far we have taken the bean as a representative flowering plant. In order, however, to have a comprehensive knowledge of the morphology and physiology of phanerogams, we must examine some structures of which the bean does not furnish examples.

111. The bean is an *annual* plant, *i.e.*, it grows for only *one season*, and hence its stem shows only one zone of fibro-vascular tissue. *Perennial* exogens, however, such as trees, bushes, etc., whose vitality lasts through a series of years, repeat each season the same succession of phenomena in the growth of the stem as have been observed in the bean during the single season to which its life is limited; that is, the stem produced during the first season is, at the commencement of the second season's growth, surrounded with a zone of parenchyma; soon bundles of woody fibre and vascular tissue, in wedge-shaped masses, become developed in the midst of this parenchyma, leaving, as before, interspaces, which, as the fibro-vascular bundles become larger, are thinned out by the consequent lateral pressure into vertical plates of cellular tissue, and designated the *medullary rays*. These plates are not continuous all the way down the stem; the whole system of *medullary rays* consists of a *series* of vertical plates arranged radially from the centre, the plates of one series *not* being continuous with those next above or below them. This arrangement, besides providing for the passage of air and sap from the peripheral parts of the stem towards its interior, adds considerably to its strength and firmness.

112. It is principally through the *youngest* woody layers (the *sap-wood* or *alburnum*) that the *crude* sap *ascends* through the stem. The returned, or descending, elaborated cambium pervades all parts of the stem where the cell

walls are thin enough to absorb it. But as the woody fibres are continually adding *lignine* and mineral matters to the interior of their cell walls, the latter ultimately become impermeable to that fluid. The protoplasm disappears; the cell wall, with its hardened contents, remains as *duramen* or hard wood, while the peripheral parts (the *alburnum*) whose cells still retain their protoplasm in an active condition, and whose cell walls are still capable of absorbing the formative cambium fluid, sustain the life of the plant. Thus the interior of the stem may be hard wood, *without any vitality*, and preserved from decay simply by being shielded from external influences by the living structures surrounding it, while the youngest wood nearest to the surface is still endowed with active life, to be, in its turn, during the next season, surrounded by structures still younger. The intermediate zones are in a transitional condition, being formed of structures whose cell walls are thickening, and hence capable of absorbing only a proportionately small quantity of cambium, till they, too, in their turn, become dead duramen, or hard wood. The "cambium zone," where the fluid is found in greatest abundance, lies *outside* of the alburnum, but *inside* of the bark. Hence the cells on its inner side are differentiated into parenchyma and fibro-vascular tissue for the medullary rays and *wood* of the stem, while its outer part is, in a similar fashion, metamorphosed into the structures of the bark. The newest layers of bark are consequently on the inside, and by their own growth, as well as by the enlargement of the stem itself, they push out the external layers. The outermost layer cracks and falls off, its place being taken by the layer next adjoining.

113. The bark itself, in young plants, consists of three layers.

a. The innermost, *liber* (or *endophlœum*), consists of fine woody fibres arranged in bundles, often beautifully interlaced with each other, the intervening spaces being filled up with cellular tissue. It is from this layer of the bark that the fibres of flax, hemp, etc., are derived.

b. Outside of this is the *green* or *middle* bark (*mesophlœum*) formed *entirely* of cellular tissue. It generally contains chlorophyll, and seldom survives the first year.

c. The third or outermost is the *corky*, or *suberous* layer (*epiphlœum*), consisting of flattened cells, which are usually shed as the stem increases in thickness. This layer is well developed in the cork oak (*quercus suber*), and furnishes the substance known as *cork-wood*.

In young plants, all these layers are invested with a covering of epidermis, which has been already described.

114. The growth of the stem in endogens (figs. 57, 58), is somewhat different from that observed in exogens.

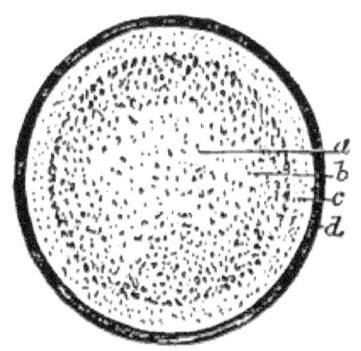

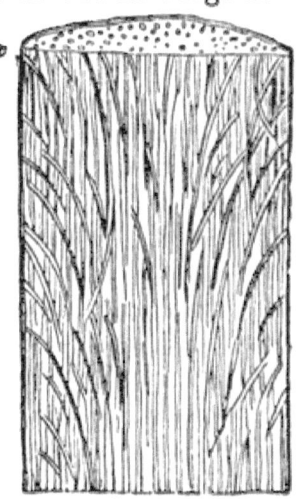

Fig. 57.—Transverse Section of an Endogenous Stem (palm). *a*, Medullary portion; the fibro-vascular bundles (represented by dots) are less numerous than at *b*, where they are crowded upon each other, forming a dense layer; *c*, zone of less compact and more slender fibres (analogous to liber); *d*, epidermis (cellular).

Fig. 58.—Vertical Section of an Endogen (palm), showing the fibro-vascular bundles, *a*, *a*, proceeding downwards (1) towards the centre; (2) towards the circumference.

The cambium fluid does not *surround* the stem, but exists in large quantity at its "growing point," where it is elaborated by the vital action of the leaves, a tuft of which is always observed growing at the summit. Hence, in the endogen, the differentiation of the cambium is not *horizontal* as in the exogen, but vertical, or downwards. From the growing point, the cells, as they become differentiated into vessels and fibres, project downwards, converging first towards the centre, but afterwards diverging outwards (fig. 58). The fibro-vascular bundles thus formed continue their course till they ultimately reach the roots, or attach themselves to the tough hardened tissue of the outer or cortical layer, a structure which corresponds to the bark in exogens, but differs from it in being inseparable from the rest of the stem, and in being harder (fig. 57, *c*). Its structure and mode of formation are also different from those observed in bark.

115. The fibro-vascular bundles, as they proceed downwards, have their structures altered considerably. After leaving the "growing point," at the base of the tuft of leaves, they consist of spiral vessels, porous vessels, a few laticiferous vessels, woody fibres, resembling those of liber, and some cellular tissue. As the bundles descend, the spiral vessels first disappear, and subsequently the porous vessels. When they reach the cortical layer, and become incorporated with it, nothing remains but fibrous tissue, which, by its division and interlacing, forms a close and tough network.

The outer part of an endogenous stem is the hardest, in consequence of its mode of formation. This, as has been already stated, is exactly the reverse of what has been observed in exogens.

116. An endogenous plant has generally several roots all attached at one point to the stem, forming what is called a *fascicle*. The structure of the root resembles that of the stem.

117. The typical leaf of an exogen has been already described. It has a lamina or blade, and a petiole or leaf-stalk. The latter commonly forms at its base an expansion, called a *vagina*, which wholly or partially embraces the stem. In grasses the entire petiole is thus

transformed. In some plants there is no petiole, and the leaf is then said to be *sessile*. The sepals, petals, anthers, and carpels of flowers have already been referred to as modified leaves. We have now to add that leaves, or portions of them, also become metamorphosed into *stipules, bracts, bud scales, spines, tendrils, phyllodes*, and *ascidia* or *pitchers*.

a. The vagina very often assumes the form of two leaf-like projections called **stipules**, at either side of the petiole. These may be free (*caulinary*), or attached to the petiole (*adnate*), or to each other (*connate*). Sometimes stipules are transformed into *spines* or thorns, as in the gooseberry, and into tendrils, as in the melon. Occasionally, as in lathyrus aphaca, they assume the functions of leaves.

b. **Bracts** are small leaflets, which are frequently observed at the bases of flower stalks. In function they are analogous to cotyledons: they serve to nurse the young flowers until they are sufficiently grown to digest and respire for themselves.

c. **Bud Scales** are modified leaves, hardened to protect the young bud from external injury.

d. **Spines.**—The *veins* in some leaves, as in the holly, become hardened at their extremity, and form *spines*. Occasionally, stipules also are metamorphosed into spines, as in the gooseberry. In the furze, some of the spines are modified leaves.

e. **Tendrils.**—In other cases, the veins (the midrib more especially) become lengthened out into tendrils, adapted to twine round a support, as in the pea. In most instances of this sort, the lamina itself becomes aborted, the tendril thus appearing as a metamorphosed *leaf*. In the cucumber tribe, the tendril is a modified stipule.

It is necessary, however, to observe, that sometimes spines and tendrils are modifications of *branches* rather than of *leaves*. The tendril of the vine is a modification of the *stem*. It is also of some importance to notice the distinction between a *spine* (or thorn) and a *prickle*. The former may be observed in the common thorn as a *woody outgrowth*, or modified branch, and the latter on the stem and midrib of the briar as a hardened *epidermic* appendage, or modified *hair*.

f. **Winged Petioles — Phyllodes.** — The petiole sometimes becomes expanded so as to become "winged," as in the orange and Venus's fly-trap (fig. 61, *p*); or a *phyllode*, as in acacias and pitcher plants (figs. 59, 60, *a*). In the former, the phyllodes are generally fitted vertically to the stem. When the lamina is small or obsolete, the phyllode discharges the functions of a leaf.

g. **Ascidia** or **Pitchers.**—The petiole occasionally, but sometimes the lamina, becomes modified into what is called an *ascidium* or *pitcher*, as in *Sarracenia*, *Nepenthes*, and *Dischidia Rafflesiana* (fig. 60).

In Venus's fly-trap (*Dionæa Muscipula*), which also has a winged petiole, the lamina becomes hollowed and secretes a viscid fluid, which has the property of detaining and dissolving insects, the secretion acting in some measure like the gastric juice of animals, to aid in the nourishment of the plants. The *pitchers*, above referred to, secrete a similar juice, and to the same end, apparently.

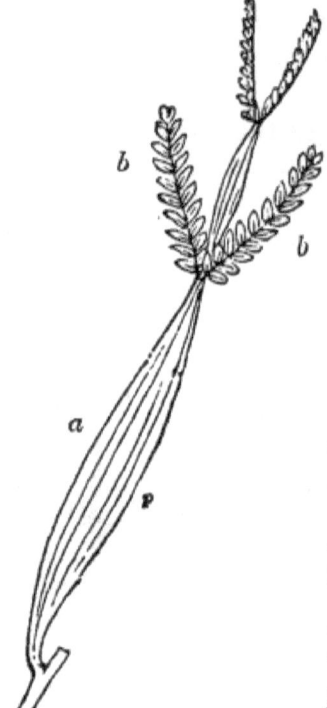

Fig. 59.—Leaf of Acacia. *a*, Petiole transformed into a phyllode; *b*, leaflets, corresponding to lamina.

118. The *venation* in the leaves of **exogens** is of that form known as *reticulated* (fig. 62, A), wherein the smaller veins, or fibro-vascular bundles, anastomose with each other, so as to form a close network. On the other hand, the veins in **endogens** (fig. 62, B), are parallel, and do *not* form a network, except in a very few plants.

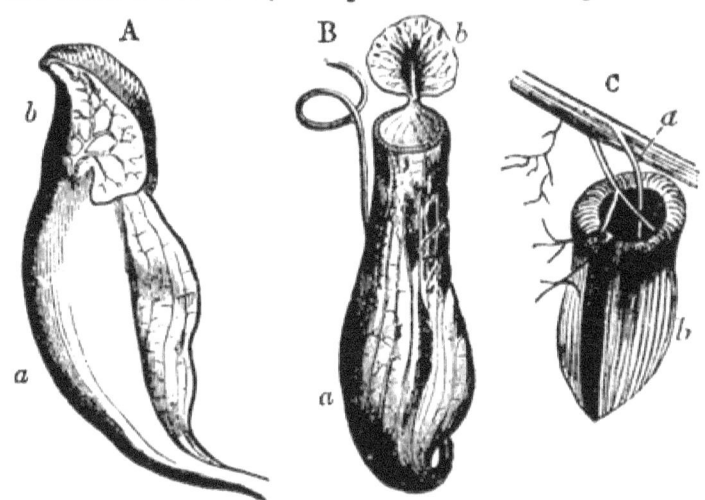

Fig. 60.—PITCHERS. A, Pitcher of Sarracenia, a Canadian marsh plant. B, Pitcher of nepenthes, a pitcher plant from the East Indian Archipelago. C, Pitcher of Dischidia Rafflesiana. In all, *a* represents the phyllode, *b* the lamina.

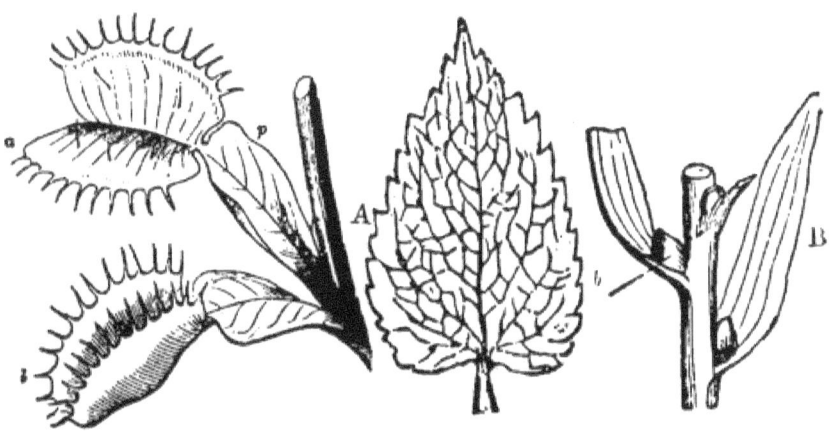

Fig. 61.—VENUS'S FLY-TRAP (*Dionæa Muscipula*), a North American marsh plant. *p*, Phyllode "wing;" *a*, lamina.

Fig. 62. — RETICULATED AND PARALLEL VENATION. A, Leaf of exogen, showing the netted veining; B, leaf of endogen, showing parallel veins.

119. In **Exogens**, the sepals, petals, and stamens are, as a rule, arranged in whorls containing five, or sometimes four, of such parts, or some multiples of these numbers—five being oftener than four, the basic number. In **endogens**, the parts of the flower are arranged in sets of three, or of some multiple of three.

120. The embryo of the **exogen** has *two cotyledons* (fig. 55) and a radicle, which is a prolongation downward of the axis. Hence the embryo is *dicotyledonous*, and the germination *exorhizal*. In the **endogens**, the embryo has but *one cotyledon*, and its radicular end remains *within* the embryo, while it gives off fibrils to form a fascicle of roots. Wherefore, in this instance, the embryo is *monocotyledonous*, and the germination *endorhizal*.

The following is a brief synopsis of the distinctive characteristics of exogens and endogens:—

In **Exogens**: (1) the wood is exogenous; (2) the veins of the leaves are netted; (3) the fructification is formed upon a quinary or quaternary type; (4) the embryo is dicotyledonous; (5) the germination is exorhizal.

In **Endogens**: (1) the wood is endogenous; (2) the leaves are straight veined; (3) the organs of fructification are ternary; (4) the embryo is monocotyledonous; (5) the germination is endorhizal.

CHAPTER XII.

THE FRESH-WATER POLYPE (HYDRA), AND THE SEA ANEMONE (ACTINIA).

121. THE amœba has been described as a type of that portion of the animal kingdom called **protozoa**. In this sub-kingdom each individual organism is composed of but one cell, formed of a jelly-like substance (sarcode), with or without a covering of harder material, such as cellulose, chitin, lime, or flint. The protozoa are either solitary and independent like the amœba, or live in colonies in a communistic fashion like the animalcules in sponges and foraminifera.

122. The **Hydra**, or **Fresh-water Polype** (figs. 63, 64), as the latter name implies, is an inhabitant of *fresh* water; but it is typical of the class **Hydrozoa**, one of the two great divisions of the sub-kingdom **Cœlenterata**, of which class most of the genera, such as jelly-fishes, sertularidæ, etc., live wholly in *salt* water. The **Actinia**, or **Sea Anemone** (fig. 69), on the other hand, is never found in fresh water. It may be regarded as the type of the second great division (called **Actinozoa**) of the same sub-kingdom, cœlenterata. Both of these typical forms resemble each other a good deal in their structure and physiology.

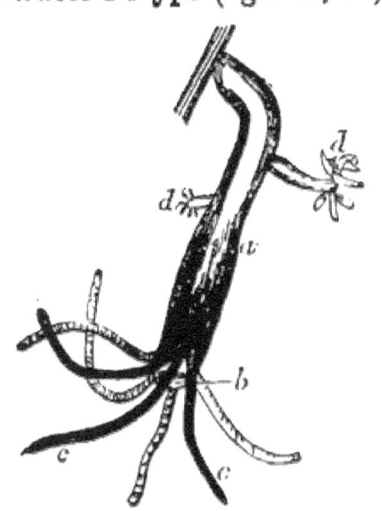

Fig. 63.—HYDRA attached by its base to a floating weed. *a*, Body; *b*, mouth; *c, c*, tentacles; *d, d*, young hydræ produced by budding.

123. The *cœlenterata* (hollow entrailed animals) are distinguished from all other animal forms which possess a distinct internal cavity, by not having an alimentary or digestive canal completely separated from, or having no direct communication with, the "somatic" cavity, *i.e.*, the general cavity of the body.* If we could imagine a dog, for instance, to live by simply digesting his food in the general body cavity, from which all, or nearly all, the viscera had been previously extracted, we should have some idea of the essential characteristic of a cœlenterate

---

* "In all the cœlenterate animals, either the general body-cavity *is* the digestive cavity, or if there be a distinct digestive tube, this opens directly into the body-cavity."—*Nicholson.*

animal. Both the hydrozoa and the actinozoa have the body wall composed of two coats, separated by a layer of muscular fibre. The outer is called the **ectoderm**, the inner the **endoderm**, and both are made up of nucleated cells. As a general rule, both classes are also provided with a circlet of tentacles ranged round the mouth for seizing their prey, and with peculiar stinging organs called "thread cells," "nematocysts," or "cnidæ." These very peculiar weapons of offence will be described further on. Of all the cœlenterata, the ctenophora alone furnish any trace of a nervous system.

124. The hydrozoa use the general cavity of the body as a digestive sac (see fig. 65), in which the food is prepared for being absorbed by the cells of the body wall. They also have the reproductive organs on the *outside*, formed from the ectoderm. On the other hand, the actinozoa have the digestive cavity *included within* the general somatic cavity, but *directly opening into it below* (see fig. 70, A). In all other animal forms possessed of a distinct digestive cavity, there is no direct communication between it and the general cavity of the body.

125. The digestive cavity in the actinozoa is kept in its place by a number of membranous partitions called the **mesenteries**, extending from it radially towards the wall of the body (see fig. 70, B). The "perivisceral cavity" is thus divided by the mesenteries into a number of chambers, which communicate with the tentacles above, and with general somatic cavity below. It is on the sides of the mesenteries that the reproductive cells are produced.

We now proceed to deal with these two typical organisms, *i.e.*, hydra and actinia, in detail.

THE HYDRA OR FRESH-WATER POLYPE.

126. This singular and very interesting animal was first observed in 1740. The Abbé Trembley, of Geneva, wrote an entire treatise upon it. It is about one-eighth of an inch in length (fig. 64). If deposited in a glass vessel containing water, it will attach itself to that side on which the light strikes. We generally find the cavity of the body filled with either a brownish or greenish matter. The former condition is met with in *hydra fusca*, the latter in *hydra viridis*. The hydra feeds on smaller animals that may come within its reach, such as the water flea, etc., by seizing them with its tentacles, transferring them to the mouth, and thence to the stomach (or general body cavity), where they undergo dissolution. The soft parts are transformed into nutriment, while the harder indigestible parts are thrown out again at the mouth.

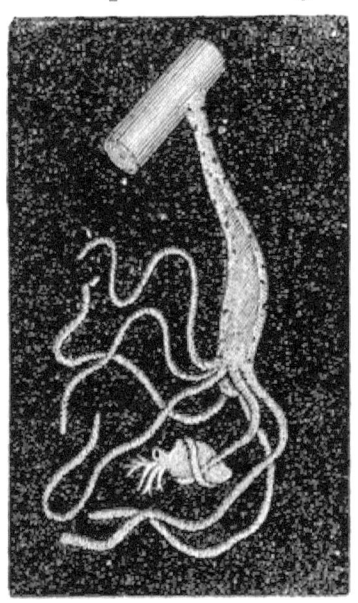

Fig. 64.—HYDRA in the act of seizing a Water Flea.

Two modes of multiplication are observable in hydra; budding or gemmation in summer, and *gamogenesis* by fertilised ova in the autumn. The subject of reproduction will be more fully treated hereafter.

Should the tentacles be touched with a needle, the *whole* animal will contract itself into a globular mass. It thus shows evidence of contractility and sensibility. Its sensitiveness to light has been already referred to.

127. Morphology of Hydra.—It is difficult to determine the true histology of a hydra, but there are certain features in its structure that are tolerably well known. The body is composed of two distinct layers of nucleated cells, the ectoderm and the endoderm, the former being the outer, the latter the inner, layer (figs. 65 and 66). This distinction of layers is found even in the tentacles, which are blind tubes projecting from the body, and arranged in a circle or "whorl" round the mouth. Their cavities, too, are continuous with that of the body.

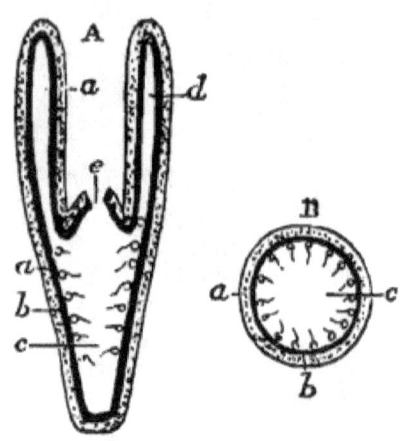

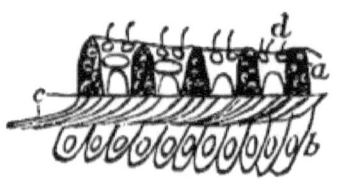

Fig. 65.—Diagrammatic Representation of Hydra. A, Longitudinal section; B, transverse section. *a*, Ectoderm (the shaded space); *b*, endoderm (the black space); *c*, "somatic" cavity, serving also as *digestive* cavity; *d*, *d*, cavities of tentacles; *e*, mouth. Cilia are represented on the interior of endoderm. (Compare with fig. 70).

Fig. 66.—Diagrammatic Section, showing the cell structure of ectoderm and endoderm, and the muscular fibres between. *a*, Cells of ectoderm; *b*, cells of endoderm; *c*, muscular fibres; *d*, thread cells. The ectoderm appears wrinkled, consisting of large and small cells ranged alternately, the interspaces being filled chiefly with "thread cells."

128. The Thread Cells (fig. 67), which abound more in the ectoderm of the tentacles than elsewhere, are very peculiar bodies. They are formed from the protoplasm of either the larger or smaller cells, and contain a long thread or filament coiled up within a tube formed by a part of the cell wall *introverted*. When the *cell is compressed*, the "introverted tube," with its attached filament, is projected outward by the simple pro-

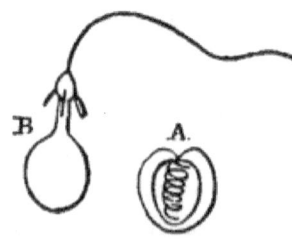

Fig. 67.—A, Cell with thread coiled up; B, cell with thread and barbs retroverted.

cess of *eversion*, just as the finger of a glove may be turned inside out. The filament is, in reality, attached to the outer side of the cell wall, and is thus an appendage *external* to the cell, though drawn inward, when coiled, by the simple involution of a part of the cell wall. The threads are suddenly ejected, as if by a spring, when the animal wishes to paralyse its prey, and render it more easy to capture. In the physalia, or " Portuguese man-of-war," the threads are a very formidable apparatus indeed, and capable of inflicting a very severe sting.

129. Between the ectoderm and endoderm there is a layer of muscular fibre, by means of which the animal is enabled to contract its body. As the hydra can also elongate itself, thus rendering itself more slender, it is supposed that there must also exist transverse fibres, though from observation alone we cannot pronounce positively as to their existence. Suppose a to be a cell of the ectoderm (fig. 68), its protoplasm manufactures a set of fibres which take the course of the lines b seen in the figure. This is presumed to be the origin of the fibro-muscular layer. It corresponds to the woody fibres in plants, being formed somewhat after the same fashion. The "thread cells" are also formed by the cells of the ectoderm, and at the expense of their protoplasm.

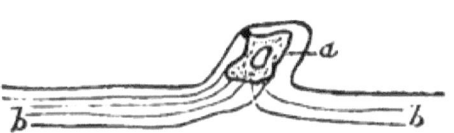

Fig. 68.—a, Cell of Ectoderm, producing b, b, muscular fibres.

130. The endoderm (fig. 66, b) is made up of nucleated cells, larger and longer than those in ectoderm. Most of them have a large vacuole, and are interspersed with granular bodies and thread cells as in ectoderm. The endoderm is richly supplied with cilia.

131. **Physiology of Hydra: Alimentation.**—We have observed that the hydra can seize upon animalcules floating within reach of it, convey them to its mouth, digest

them in its body cavity, and assimilate to its own substance the nutritive materials. In other words, the hydra abstracts the protein parts of its food, and returns the remainder to the water. It is not well understood *how* this is effected. It is supposed that the endoderm secretes a fluid analogous to the gastric juice, which acts chemically upon the substances taken in as food, and prepares them for absorption into the general system.

132. But how is this nutritive material, the result of gastric digestion, distributed to the general system? The cilia of the endodermic cells keep up a constant current of the nutritive fluid of the interior. Furthermore, every movement of the animal also promotes this circulation. By means of the longitudinally-arranged muscular fibres, the hydra can shorten itself almost indefinitely, and, by doing so, cause a movement of the fluid through the cavity and into the tentacles, while by contracting the transverse fibres it becomes narrowed and elongated, and a new current is originated. So also the tentacles may be extended by pushing some of the fluid into their cavity. All this is an approximation to the circulation of the blood in the higher animals. The ultimate nutrition of the hydra arises from the imbibition, by the cells, of the nutrient fluid by which they are surrounded, and from its passage where necessary from cell to cell, just as in chara. The successive contractions of the tissues imply waste; hence free oxygen must be present in the water to oxidate them thoroughly. Carbonic acid is disengaged, as well as some nitrogenous matters, also oxidated, in some such form as urea, or uric acid.

133. As there are no special organs for digestion or circulation, so also there is no special excretory apparatus. The carbonic acid formed by oxidation of the tissues must be given off into the liquid which surrounds the cells. Each cell thus excretes its own effete matters. We know that as oxidation produces heat, so the animal must give out heat in some degree, as even plants do in what is termed their respiration, when they absorb oxygen and

evolve carbonic acid. Thus, in the hydra there is no differentiation of parts—no specialised apparatus such as we find in the higher animals.

134. In further examining the hydra, we observe that it possesses the property of irritability; in other words, a stimulus applied externally will cause it to contract. The general facts connected with the irritability of the hydra may be thus presented:—

(*a.*) There is observed a contraction of some one or any part of the body, independently of any external stimulating cause.

(*b.*) The contraction also follows upon a stimulus applied externally.

(*c.*) The contraction is not confined to the part acted upon by the stimulus.

(*d.*) All the movements are co-ordinated to each other, and subordinated to a general end or purpose.

135. (*a.*) **Locomotion** in hydra is effected in various ways. Sometimes the animal fastens itself by its mouth, or by some of its tentacles, to a fixed object; it then detaches itself at the other extremity (the foot), which it gradually moves to a point close to where it is attached by the head. It next fixes itself by the disc on the foot, extends its body, and a second time holds fast by the anterior extremity, after which it draws up the hinder as before, thus progressing by a "looping" sort of movement. This may be regarded as the *walking* gait of the hydra. When it wishes to *run*, the mode of locomotion is somewhat different. Instead of fixing the foot near the point where the body is supported by the tentacles, it swings the body over the head, and fixes the foot as far as possible in advance. The head is now in its turn swung over the foot, and fixed at a point still further forward. Thus it progresses, like an acrobat, by a succession of somersaults.

Occasionally it projects the flat part of the foot above the water, where it very soon becomes dry. This now forms a sort of float, which enables the animal to work itself forward by the movement of its tentacles, or to take advantage of the wind, or of the currents of the water, to have itself transported to some distance from its previous resting place.

The mechanism by which all these various movements are effected must be somewhat complicated. But they are all intelligible if we suppose the existence of two sets of muscular fibres—one longitudinal, the other transversal. The longitudinal fibres will enable the animal to shorten itself, or to turn to one side or the other; while the transverse fibres, by compressing the fluid contained in the cavity of the body, or in those of the tentacles (thus narrowing their calibre), will lengthen out the body or the tentacles to any required extent.

136. (*b*.) It is not easy to understand how the stimulus created by the action of a needle in pricking a cell is conveyed to the muscular fibres, seeing there are no *nerves*; but if we suppose the protoplasm of the cell to contain nerve matter *not differentiated*, the explanation will be easy. The cell possesses the *property* of nervous matter, inasmuch as it is capable of conveying a stimulus to a muscular fibre. Nevertheless, it would not be correct to call these cells *nerves*. We have here nerve and muscle in their least differentiated condition.

137. (*c*.) It is an interesting question to consider why the contraction is not confined to the part to which the stimulus is applied. May the stimulus be communicated from cell to cell? (*d*.) How again is it that the various movements are co-ordinated to each other? In the higher animals the answer to such questions as these is comparatively easy, all the movements of the body being governed and directed by the nervous system. But there is no nervous system in the hydra. All we can say in reference to these questions in the case of the hydra is, that a *cell can convey a stimulus*.

138. If a hydra be cut in pieces, new hydræ will be formed of the several parts. It may even be turned inside out, like the finger of a glove, and it will continue

to enjoy life as before, without seemingly suffering from the injury. When a hydra has its body turned inside out, the tentacles will spontaneously turn themselves in a similar fashion, adapting themselves to the altered circumstances. Hydræ must, therefore, be very tenacious of life.

139. Reproduction.—There are two modes of multiplication observed in hydra (*a*, asexual; *b*, sexual) corresponding to the reproduction of plants by buds and by seeds.

140. (*a.*) The asexual mode of propagation in hydra is by *budding* or *gemmation*. A pouch-like projection, of both the ectoderm and endoderm, is first observed on the side of the animal, like an incipient tentacle, and like it, too, enclosing a recess of the body cavity. An opening at the extremity to form a mouth, and the growth of a bunch of tentacles around it, soon after give the bud the appearance of a young hydra attached by the base to its parent (see fig. 63). In this condition the offspring, though it partakes of the nutritive fluid of the parent, yet manages to capture food on its own behalf, and may even bud out a hydra of the third generation. After some time, however, a constriction appears at the point where it is attached to the parent. The attachment at length gives way, and the bud drops off as an independent hydra. As many as half a dozen hydræ may be observed thus growing together, and a single generation of independent hydræ may, under favourable circumstances, be thrown off within the space of twenty-four hours. This mode of reproduction is called "discontinuous gemmation." There is no limit to its operation when favoured by warmth.

In most of the marine hydrozoa, the animalcules thus budded off remain permanently attached to the parent, either by themselves or by their hardened skeleton; and as the operation is repeated indefinitely, a colony of hydroid polypes is formed, resembling a plant with its stem and branches. This form of propagation is called "continuous gemmation." There are corresponding differences in the agamogenesis of phanerogams and ferns—the former, as a rule, showing *continuous* gemmation by *buds*, the latter, *discontinuous* gemmation by *spores*.

141. (*b*.) When winter approaches there is also sexual multiplication. The sperm corpuscles or male elements are produced near the tentacles; the germ cells farther back. The former are developed in little swellings, occasioned by the special development, in particular patches, of the ectoderm. These consist of masses of cells, which, when ripe, burst the enclosing membrane, and escape into the water. Each cell contains a little body provided with cilia. It is a sperm corpuscle,* corresponding to, and very much resembling, the antherozoid of the vegetable kingdom. When floating in the water the sperm corpuscles burst their cell wall, and by means of their cilia swim about till they reach the germ cells (which, in the animal kingdom, are called **ova**—each an **ovum**). They fertilise the ova by entering into their substance, thus blending the two protoplasms.

The *ova* or germ cells are also a special development of the cells of the ectoderm, which, after fertilization, escape likewise from the enclosing membrane into the water. Their protoplasm takes somewhat of a stellate form, and encloses a nucleus answering to the germinal spot in the ova of the higher animals.

After fertilization, however, the germinal spot disappears, the contents of the ovum divide and subdivide until there is nought left but a granular mass, each granule being a minute nucleated cell. While still undetached from the body of the animal, the ovum encloses itself in a covering of *chitin*, produced by the action of the protoplasm. When the coat is fully formed, the ovum drops off from the body of the hydra, and falls into the mud, where it passes the winter, and remains till circumstances favourable to its development arouse it again into activity.

* In works on physiology the sperm corpuscle in animals goes under the name of *spermatozoön*.

The enclosed animalcule is then found to consist of two layers of nucleated cells (corresponding to ectoderm and endoderm). It bursts its chitinous covering, comes out ciliated, swims about for some little time, settles down and attaches itself to a fixed object, gives off projections to form the tentacles, and grows up into a hydra.

THE ACTINIA, OR SEA ANEMONE.

142. The **Sea Anemone** may easily be noticed by visitors to the sea-shore. It may be observed between high and low water adhering to rocks and stones, often in little pools left by the retreating tide.

Fig. 69.—A, Actinia; B, same, with tentacles retracted. *a*, Body or "column;" *b*, tentacles; *c*, mouth; *d*, disc; *e*, foot.

143. Morphology of Actinia.—In form it is like a truncated cone (fig. 69) securely fastened by its base to the rock. At its upper end is an elliptical cavity which forms the mouth: round this is a flat disc which bears upon its margin a large number of tentacles alternately arranged in concentric circles like the whorls of a flower. The body—or "column," as it is called—is of a soft, leathery consistence. A vertical section of the actinia (fig 70, A) shows that its digestive cavity terminates about half-way down the trunk, opening below by a wide aperture into the general body cavity. A number of membranous partitions, called the **mesenteries**, arranged vertically, and passing radially from the digestive sac within to the body wall without, divide the "perivisceral cavity" into a number of chambers. Fig. 70, B, is a diagrammatic transverse section of the trunk, showing the disposition of the digestive cavity d, the body wall k, l, and the mesenteries f. As in the hydra, the body wall consists of two coats, an ectoderm k, and endoderm l, with a layer of muscular tissue — the latter very highly developed.

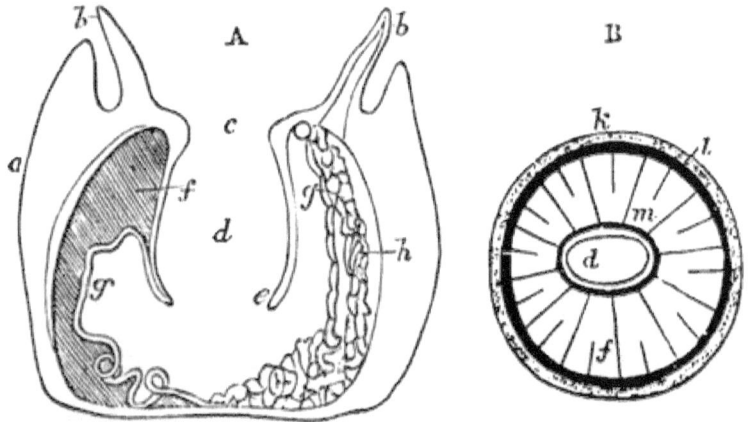

Fig. 70.—DIAGRAMMATIC REPRESENTATION OF ACTINIA. A, Vertical section; B, transverse section. a, Body or column; b, b, tentacles; c, mouth; d, digestive cavity; e, its side wall; f, mesentery; g, *craspedum*, containing thread cells; h, bands which develop either ova or sperm corpuscles. B, k, ectoderm; l, endoderm; m, ovarian or mesenteric chamber. In A, the plane of section is supposed to coincide with a mesentery at one side, and to divide a chamber at the other. (Compare with fig. 65.)

Cilia abound in the side walls of all the cavities. The tentacles are hollow tubes opening into the perivisceral chambers, and formed also of ectoderm and endoderm, with muscular tissue. They have an aperture at the point, which the animal can close at will. It can also close its mouth and retract its tentacles, like the hydra, when acted upon by an external stimulus, and assume the appearance of a roundish pulpy mass (fig. 69, B). Like the hydra, too, it is exceedingly sensitive to light. "When fully expanded, and displaying their glowing colours to the mid-day sun, a passing cloud will cause them to fold in their flower-like summits; and even the shadow of the hand will produce the same effect."* The actinia possesses great muscular strength. One was made to raise a weight of six pounds. It habitually seizes, as its ordinary prey, shrimps, prawns, and the smaller crabs and fishes, etc., which require for their capture considerable prehensile power. In this it is assisted by the benumbing action of the "thread cells" with which the tentacles are largely furnished. These cells are, in fact, so numerous, that it would seem the tentacles are composed of but little else. At the lower free edges of the mesenteries are a set of curiously-twisted cords, called *craspeda* (fig. 70, A, *g*). Their use is unknown, but they are abundantly supplied with thread cells. It is supposed that they are sometimes extended through the circular apertures occasionally met with in the body wall. Attached to the faces of the mesenteries are red bands which develop either sperm cells or ova, according to the sex of the animal; for the actinia, unlike the hydra, is bi-sexual. In many others, however, of the actinozoa both sexes are frequently combined in the same individual.

144. Physiology of Actinia.—The stomach of the actinia has considerable digestive power. Not only are the soft parts of the prey dissolved out by the gastric fluid, but even the tendons and ligaments are thus converted into nutriment, nothing being rejected as indigestible but the shells or bones. When it is not convenient to return large shells, etc., by way of the mouth, they are thrust out through an extemporised opening at the side of the body; after which the wound soon closes by healing. In ejecting indigestible matters by the mouth, the stomach is raised upwards, and sometimes so far as to be completely everted, and protruded beyond the mouth, in this manner getting rid of its contents. It is not necessary

* Carpenter.

that the whole of the prey should be included within the stomach during digestion, for one part may be undergoing dissolution below, while the remainder projects out of the mouth, waiting its turn. The nutrient fluid, the result of gastric digestion, passes into the general body cavity, whence it is conveyed by ciliary and muscular action into the perivisceral chambers, and through them into the cavities of the tentacles. This approach to a circulation is similar to what has been observed in hydra.

145. Like the hydra, the actinia is not known to possess a *nervous system*.* Like it, too, it not only shows a high degree of contractility, but gives evidence of irritability as well, when acted upon by a stimulus applied externally. It has been already observed that it is highly sensitive to light. It also manifests a notable sensibility of changes in the condition of the atmosphere, so much so that the Abbé Dicquemare regarded it as a valuable marine barometer—its expansion indicating fair weather, its closure, bad weather, and its extreme contraction, boisterous or stormy weather.

146. Locomotion is effected as in the snail by the alternate contraction and expansion of its "foot." It often inflates itself with water so as very considerably to enlarge its size, and become of nearly the same specific gravity as the surrounding medium. It then floats about till it finds a convenient resting place. While thus supported by the water, it can also, with its mouth turned downwards, *walk* along the sea bottom by means of its tentacles. Sometimes, however, the actinia thus distends itself without removing from its position. In such cases the extra quantity of water is supposed to be taken in for respiratory purposes.

147. The actinia can bear mutilation with as much im-

---

* "A nervous system has not yet been proved to exist in any of the actinozoa, except in the ctenophora, and in none are there any traces of a vascular system. Some actiniæ are said to have short optic nerves distributed to the pigment-masses at the bases of the tentacles, and these masses possess crystalline lenses."—*Nicholson*.

punity as the hydra. It may thus be artificially multiplied by fission—a process which it also undergoes naturally. It may be frozen and thawed again, or may be immersed in *salt* water heated to 140° F. without any apparent injury to its health. It cannot, however, live in *fresh* water, whether cold or hot.

148. Reproduction.—Asexual multiplication, when it occurs in actinia, is always by *fission*—one polype breaking up into two or more. *Budding* is a form of propagation not observed in actinia, but is common in some others of the actinozoa.

149. Sexual reproduction takes place as in hydra; but in actinia the sexes are found in separate individuals. The ova and sperm cells are developed in the interior, and the young are usually not liberated until they attain a comparatively high degree of development. Occasionally, after escaping from the ova, the gemmules are passed out, ciliated, at the apertures in the tentacles; but, generally, the young are retained in the ovarian chambers till they assume almost the appearance of the adult animal. They then pass into the general body cavity, and are ejected through the mouth. "The young are frequently disgorged along with the half-digested food; thirty-eight appearing thus in various states of development in a single litter."

150. The *coralligena*, or coral producing polypes, do not *structurally* differ much from the actinia. Their chief distinguishing features consist—

(*a*.) In the former remaining attached to the parent organism, and to each other, after *fission* or *gemmation*.

(*b*.) In many of them being *hermaphrodite, i.e.*, in having sperm cells and ova produced in the same individual polype.

(*c*.) In having the mesenteries and the walls of the body and stomach *calcified, i.e.*, turned into carbonate of lime, by the secretive power of the cells constituting those parts of the body. The skeleton thus formed remains after the soft parts of the animal which produced it have died and decayed; and to the accumulation of these imperishable portions of a once living organism we owe not only the existence of the immense *coral reefs* now in course of formation in different parts of the ocean, but also of the vast beds of limstone rock now buried deeply in the bowels of the earth, or rising into mountain masses many thousands of feet above its ordinary level.

CHAPTER XIII.

THE FRESH-WATER MUSSEL (ANODON).

151. In passing from actinia to anodon, we make a vast stride in the animal scale; but space will not permit of our entering upon the consideration of beings occupy-

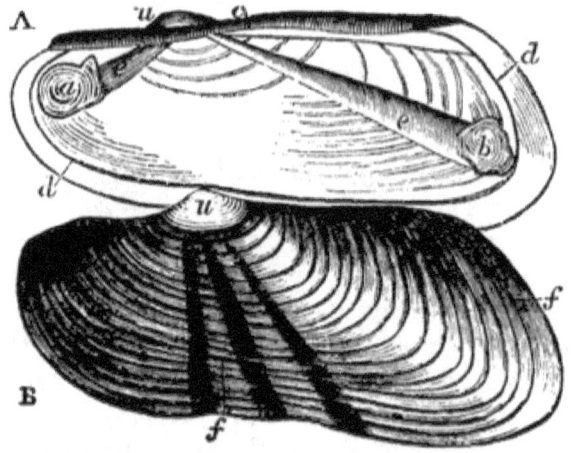

Fig. 71.—A, B, Right and left valves of anodon, the former showing the inside, the latter the outside of the shell. *a, b,* "Adductor impression" or hollow, showing attachment of adductor muscles; *c,* hinge; *d, d,* "pallial line" or "impression," a hollow curve line marking the attachment of edge of mantle to shell; *e, e,* furrows, showing the passages of the adductor muscles during growth of shell; *f, f,* "lines of growth;" *u,* umbo.

ing the intermediate stages. The **anodon** belongs to the sub-kingdom *mollusca*, and to the class *lamellibranchiata*, characterised by being devoid of a head (hence called *acephala*), and by having *gills* formed of *lamellæ*, or little sheets or leaves. As the name **Fresh-water Mussel** implies, it is found in rivers and lakes, generally buried in the mud. Most molluscs are, however, inhabitants of *salt* water.

152. **Morphology.**—When we examine the anodon, we find that it is made up of several distinct parts, each of which has its own function to perform, though they are all co-ordinated with each other, and in their totality make up the structure and life of the animal. We have, on the outside, what is usually termed the *shell* or the *valves*—by naturalists the **exo-skeleton** or outer hard part. The valves (fig. 71), which are arranged right and left, and hinged over the back by a very elastic ligament, may, at the will of the animal, be kept tightly closed by means of two strong muscles, passing directly through the body from one valve to the other. One of these muscles, the **anterior adductor**, is just *above* the mouth; the other, the **posterior adductor**, just *under* the **anus**, or terminal aperture of the intestinal canal. When we open the valves we may observe what are termed the **mantle** or **pallium** (fig. 72, *a*, *b*), and the **foot**, *d*. If we remove one side of the pallium we observe a soft gelatinous body—the body of the animal. We find that this body is enveloped by the mantle, a membranous covering, which is continuous over the back of the animal (the part nearest the ligament or "hinge"), but separated in front. Each side or half of the mantle is termed a **lobe**. The cavity *e*, *e*, between the mantle lobes in front, is called the **pallial** or **mantle cavity**. The two lobes are, however, attached, like a "buttoned coat," at a point between the **cloacal** * and mantle cavities, between the letters *e*

---

* The *cloaca* is a chamber at the hinder end of the body, into which the anus opens. It also receives the water of the epi-branchial chamber, which has been rendered impure by its

and *f* in the figure. Anterior to that point they remain separate.

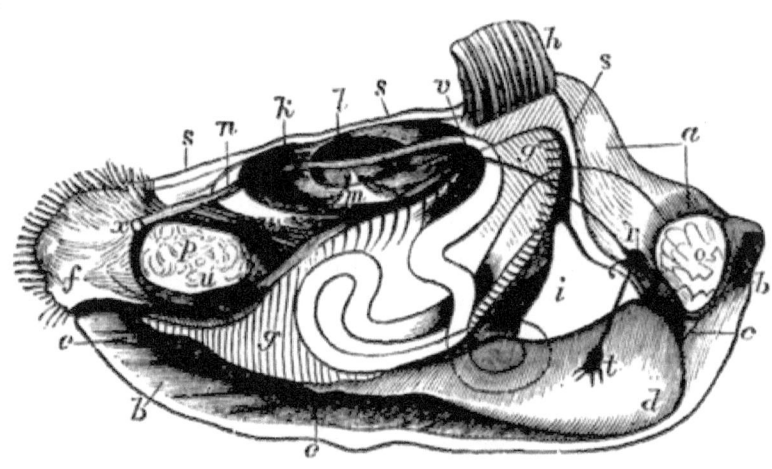

Fig. 72.—DIAGRAMMATIC SECTION OF ANODON. The right lobe of the mantle, except the portion *a*, is cut away along the line *s, s, s*; *b, b*, left lobe of mantle; *c*, mouth; *d*, foot; *e, e*, branchial or pallial cavity, enclosed by the mantle; *f*, anal or "cloacal" chamber; *g, g*, right interior gill, partially cut away; *h*, a small portion of right exterior gill, turned back; *i, i*, labial palps; *k*, pericardium; *l*, ventricle of heart; *m*, one of the auricles; *n*, rectum, after passing through heart; *o*, anterior, *p*, posterior adductor muscle; *r, t, u*, cerebral, pedal, and parieto-splanchnic ganglia, with their commissures; *v*, opening into the organ of Bojanus; *w*, one of the two posterior retractor muscles, attached behind to the posterior adductor. The intestine is supposed to be seen extending from the mouth (behind which there is an enlargement, constituting the stomach) through the substance of the foot, where it makes many windings, then through the pericardium and ventricle of the heart, and finally terminating in the cloacal chamber at *x*, the anus (see Art. 159).

153. If we put the animal, with the dorsal side upwards, in water containing some few minute grains of indigo, we shall observe a current of the fluid containing the coloured granules entering between the mantle lobes below, and again passing out by the cloacal aperture.

passage through the gills, or through the organ of Bojanus. It thus gives exit to all the waste products of the body.

The *foot*, *d*, lies in the middle line, and the four gills form a complete partition (but perforated by minute pores) between the branchial and epibranchial chambers. The **palpi** serve as guards or tentacles to the mouth; they arrest all the food particles driven forward by the current generated in the mantle cavity.

On the inner side of the attachment of the inner gill, there is an aperture called the orifice of the **organ of Bojanus**, or *renal aperture, v.* There is one on each side of the body. Indeed the whole body is bilaterally symmetrical. A line drawn from the mouth *c* to the anus *x* will represent the axis of the body; it runs *beneath* the anterior adductor, *above* the posterior adductor, and *between* the two retractor muscles of the foot. The heart and great vessels are situated dorsally as regards the axis, while all the ganglia of nervous matter—the cerebral ganglia excepted—are on the opposite or ventral side of the axis. Hence the dorsal side is designated the *hæmal*, and the ventral side the *neural*, aspect of the body. The two cerebral ganglia are, however, *above* the mouth, and the two *commissural cords*, which connect them with each other, encircle the œsophagus. This loop of nerve matter is hence called the **œsophageal collar**, and is very generally to be met with in invertebrate animals possessed of a nervous system.

154. The **gills** are each divided into two laminæ (fig. 76). The attachment of the outer lamina of outer gill being upon the pallial lobe, and that of the inner lamina of inner gill upon the foot. The **branchial** chamber A, and **epibranchial** B, are each divided into two parts by the foot *k*. The branchial chamber communicates with the mouth (fig. 72, *c*), the epibranchial with the cloacal cavity (fig. 72, *f*). The water in the branchial chamber is driven through minute apertures in the partition between it and the epibranchial chamber, and must leave the body by the cloaca.

155. This circulation of water is important to the life of the animal, for it not only supplies oxygen to the gills, but also sweeps forward to the mouth the food materials floating in it, and which have been prevented from entering the epibranchial cavity by the sieve-like structure of the gills. The labial palpi assist in gathering in the food by their ciliary action. The mouth, and indeed the whole alimentary canal, is also richly ciliated, and thus the food is quietly conveyed from the mouth to the stomach, and thence through the intestines.

156. In the anodon we do not observe any segmentation, or any appendages, such as limbs, so characteristic of the lobster and other arthropoda. There is, however, a perfect bilateral symmetry, the auricles, gills, lobes, valves, and most of the orifices having corresponding parts on both sides. There are no bones to form an endo-skeleton, but there exists an exo-skeleton in the form of two shells (or valves), formed essentially of carbonate of lime. The outside cells of the pallium secrete the matter of the shell. The outer part of this shelly substance—the cuticular layer—is formed of epidermic cells impregnated with the same mineral material. The middle layer consists of a series of vertical prisms placed side by side, and showing a laminated structure; and the third, or innermost, layer is *nacre*, formed of laminations, with fine tubuli passing into its substance. The cross section of the latter gives the appearance as if it were formed of cells like epidermis, which have become hardened in process of growth; but this is not the true cause of the phenomenon.

157. There are two valves, a right and a left (see fig. 71). The **umbo** always lies close to the dorsal margin, and is turned in the direction of the mouth. The **ligament**, which is always found behind the umbo, is simply the non-calcified shell, consisting as it does of horny or epidermic matter; so that in real strictness the shell is not bivalve at all.[*] It would be more correct to say that it was one shell with two lobes. The ligament acts an important part. It is highly elastic, and is kept on the stretch, when the shell is closed (fig. 73, A), by the contraction of the **adductor muscles** *m*. When these latter are cut, or when they become decayed, the valves gape, as the elastic force of the ligament is sufficient to open them. The shell grows by the addition of fresh layers to the interior; but these layers as they are produced extend beyond the margins of the older ones, as is seen in fig. 73, B, at *a*, *b*, *c*, *d*, *e*.

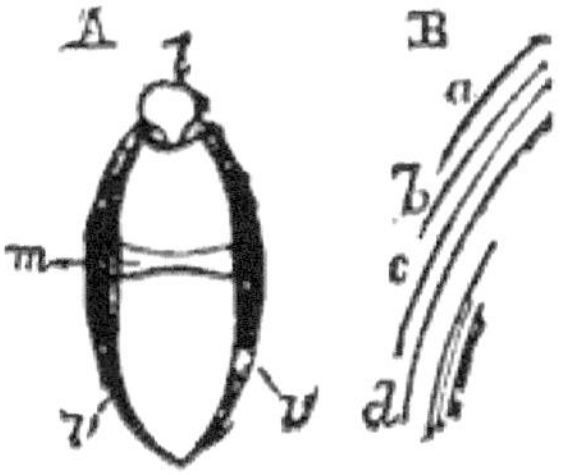

Fig. 73.—A, Cross section of closed valves. *l*, Ligament; *m*, adductor muscle; *v, v*, valves. B, Diagram showing mode of growth in shell; *a, b, c, d*, successive layers secreted by pallium.

158. A depression (fig. 71, *d, d*), called the **pallial impression**, is observed in a curving line, parallel to the margin of the shell. This is caused by the attachment of the pallium along that line. There may also be noticed two triangular furrows (fig. 71, *e, e*) extending from the umbo to the adductor muscles. These furrows are caused by the passage of each muscle in its growth from point to point, in order to adapt itself to the increasing size of the shell.

There should also be pallial markings parallel with the margin of the shell, and extending along the surface from the umbo towards the edge; but they are comparatively faint. The hollows marking the points of attachment of the adductor muscles are called **adductor impressions** (fig. 71, *a, b*).

159. **Physiology.**—We now proceed to consider the different organs of the anodon; and we shall first examine the alimentary canal. It commences at the mouth (see fig. 72), and proceeds to the **stomach** by the œsophagus. The stomach is a bag enclosed in the substance of the liver. The first turn of the intestine is downwards, or towards the *neural* side. It subsequently makes many winds and turns in the substance of the foot. At length it enters the pericardium, and passes right through the ventricle of the heart, though without communicating with

it. It finally terminates in the cloacal chamber, at the anus x. There is no natural line of demarcation between the *rectum* and the rest of the alimentary canal. The whole canal, including, of course, the stomach, is lined with ciliated epithelium, which keeps the fluids passing through it in constant motion.

160. The **liver** invests the stomach. It is supplied abundantly with blood, from which a fluid is extracted by blind *tubuli* (fig. 74, A, B) abounding in its substance. This fluid is poured into the stomach and aids in the digestion. Its secretion resembles that of *bile* in the higher animals. The epithelial cells in these tubuli are filled with brown granules, in which the secreting power possibly resides. The minute animals and plants that constitute the food of the anodon are carried by ciliary currents into the mouth, whence they pass by the œsophagus into the stomach. The food is there digested, and the resulting fluid is carried along by the currents created by the cilia of the epithelium. When ready for being assimilated, it permeates the walls of the intestines, and finds its way into the current of the blood, and thence into the tissues.

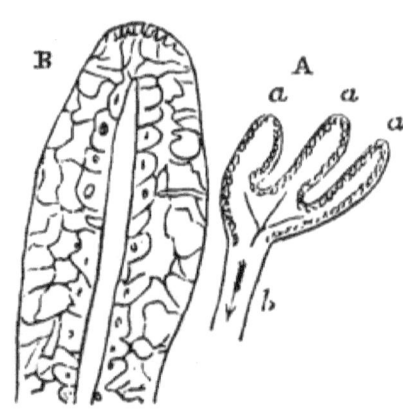

Fig. 74.—A, Three secreting tubules of liver, opening into a common duct b. B, One of the tubules highly magnified.

161. **Blood Circulation.**—In addition to the digestive apparatus, an organ, or means for *distributing* the nutritive materials to the parts of the body where they are wanted, is also necessary. In the anodon, the apparatus for effecting this is exceedingly complex. The heart (fig. 75, b, c) consists of two **auricles** and one **ventricle**, all of which are enclosed in the **pericardium** (A, p). When viewed from above, the ventricle of the heart is seen to

taper to a point where it embraces the alimentary canal. The latter, indeed, runs right through the pericardium and ventricle of the heart, and between the two retractor muscles of the foot. The heart thus acts the part of a *perivisceral chamber;* and the movements of the muscles of the foot, by pressing the walls of the alimentary canal, favour the absorption of the nutritive fluid.

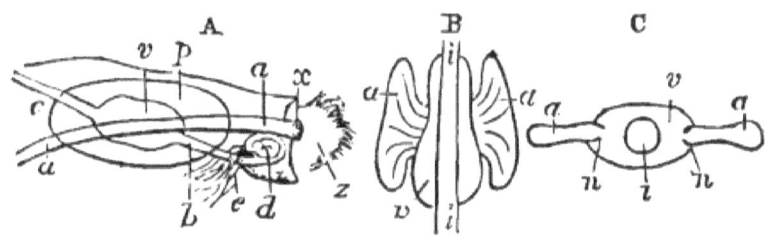

Fig. 75.— A, Side view of heart and pericardium: *p*, pericardium; *v*, ventricle; *a, a*, alimentary canal; *x*, anus; *z*, cloaca; *c*, anterior aorta; *b*, posterior do.; *d*, posterior adductor; *e*, posterior retractor. B, Heart viewed from above: *a, a*, auricles; *v*, ventricle (the auricles are represented as being pulled out; in nature they lie close to the ventricle); *i*, intestinal canal. C, Cross section of heart: letters same as in B, except *n, n*, valves of auricles.

162. When the blood passes into the auricles, the latter, by their contraction, force it into the ventricle. The ventricle then contracts, not all at the same time, as in the higher animals, but by a peristaltic movement. The blood is thus forced into both arteries, the anterior and posterior aortas (fig. 75, A, *c, b*). None can pass out into the auricles, for the valves prevent its return (C, *n, n*). The walls of the ventricle are lined with epithelium. The middle layer between the two cellular layers is made up of muscular fibre. The rythmical action of the heart is due to a sub-nervous apparatus embedded in its structure. It is a question whether the circulating *vessels* have membranous coats as in higher animals, or whether they be merely intercellular canals excavated in the tissues.

163. The **vena cava**, or venous sinus (fig. 76, *h*), is a space or lacuna hollowed out of the tissues at the lower side of the pericardium. What may be termed *venous* blood collects in it, and then passing into the *wall tissues* of the lower chamber or "glandular sac," *f*, of the organ of Bojanus, it circulates among its minute capillary and granular structures. In its passage it loses its oxidated nitrogenous products, through the action of the glandular tissues of that organ—becomes strained, so to speak, just in the same way as the blood passing through the kidneys of the higher animals is acted upon. The waste products are carried off by the water which always fills the central canal of the organ of Bojanus. Having passed out of this organ, the blood circulates amongst the vessels of the gills, where it becomes oxygenated by its exposure to the water. After oxygenation it passes into the auricles, thence into the ventricle, which forces it into the two aortas, by which means it is at once distributed to the anterior and posterior muscles, to the liver, and the other parts of the body. The blood circulating in the mantle, being more or less exposed to the action of the oxygen contained in a free state in the water, is thereby "*arterialised*" to a certain extent. This blood is not returned directly to the auricles, but to the pericardium.

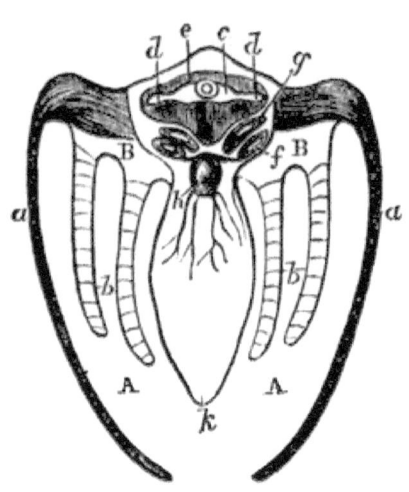

Fig. 76.—Diagrammatic Transverse Section of Anodon, through the heart. *a, a*, Lobes of mantle; *b, b*, gills, showing transverse partitions; *c*, ventricle of heart; *d, d*, auricles; *e*, pericardium; *f*, glandular sac of organ of Bojanus; *g*, vestibule, or middle sac; *h*, venous sinus; *k*, foot. A, A, Branchial or pallial chamber; B, B, epibranchial chamber, communicating with cloaca.

164. The pericardium differs considerably from the

PERICARDIUM OF ANODON. 107

organ getting the same name in vertebrate animals. In the latter it is simply a serous sac, containing a fluid which renders the motions of the heart smoother and easier; but in the anodon the pericardium contains some blood—as is proved by the colourless corpuscles (fig. 77) found in it—and affords a means of allowing the blood system to communicate directly with the exterior.

Fig. 77.—A, Blood corpuscles (colourless, amœbiform) in blood of anodon, taken from pericardium. B, same, treated with acetic acid; nucleus distinct.

In all molluscs, the pericardium communicates with the exterior directly, and forms a reservoir both for blood and for water. Near

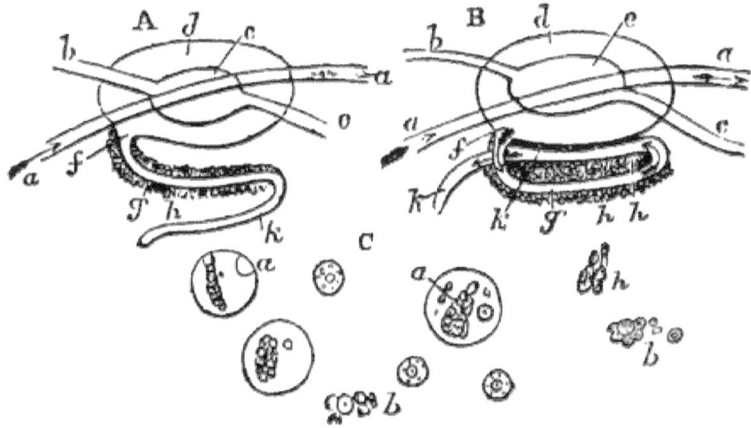

Fig. 78.—A and B, Sections of pericardium and organ of Bojanus, showing how they communicate. A gives the arrangement of the latter organ much simplified; B, as it is found in nature. a, a, Alimentary canal; b, c, anterior and posterior aortas; d, pericardium; e, ventricle; f, opening of pericardium into glandular sac, g; h, granular matter in walls of glandular sac; k, vestibule, or middle sac. The same letters refer to the same parts in both diagrams. C, Cells in glandular sac of organ of Bojanus: a, granular matter in protoplasm; b, granular matter escaped from cell.

the point where the intestine enters it, there is an orifice (fig. 78, A, B, *f*) opening into the glandular sac of the organ of Bojanus, which sac has a canal running through it, and continued through the vestibule or middle sac, *k*, until it opens finally into the epibranchial chamber.

165. Functions of the Organ of Bojanus.—The venous blood, in passing through the organ of Bojanus, loses its waste nitrogenous products, which are, as it were, *washed out* by the water passing through the *canal* in the glandular sac (fig. 78, *g*). The organ of Bojanus thus corresponds to the kidneys in function. As to the products carried outward by the water, the results of chemical analysis are conflicting. The blood in anodon contains plasma, fibrin, earthy matters, and corpuscles, but only of one kind—the colourless (fig. 77). It has nothing corresponding to the red corpuscles of vertebrate animals.

166. The Gills—Respiration. — Each gill forms two plates, connected below like two leaves in a sheet of paper. It thus forms a kind of pouch, but with transverse partitions to keep them from either collapsing or separating too much from each other (fig. 76, *b*).

The gill structures are not quite continuous, but are made up of rods (fig. 79, *a*, *a*), or thin narrow plates, furnished with ciliated epithelium cells. The cilia keep up a constant current of water between the plates, which currents, passing over the network *c, c,* in which the blood circulates, aërate the latter. The rods or bands which go to form the gills are rendered firm by bars of chitin *b*, *b*, inlaid in a regular determinate order.

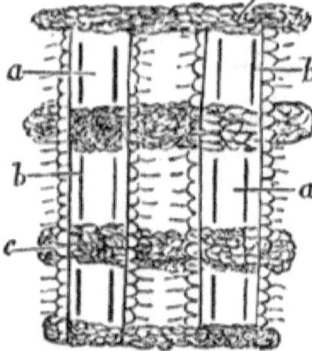

Fig. 79.—Small portion of gill magnified, to show *a, a,* tubular rods, strengthened by *b, b,* bars of chitin, and supporting *c, c,* network of capillaries. The rods are richly supplied with ciliated cells.

167. Locomotion and other movements are effected by

the *adductor*, *retractor*, and *pallial* muscles. As has been already observed, the valves are kept closed by means of the adductor muscles. The retractor muscles (fig. 80, c, d), have their origin* in those adductor muscles, and their insertion in the foot. The accompanying figure will explain their mode of action. When the *posterior* retractor muscles contract, the water is pushed forward into the foot. The fore part of the latter accordingly advances a little, and, becoming a fixed point, the hinder part is drawn up after it. This is effected (1) by the *anterior* retractor muscles driving back the water by pressing the fore part of the foot; and (2) by the action of a muscle along the base or "sole" of the foot, called the *protractor* muscle. The *protractor pedis* has its *origin* in the fore part of the foot, and its *insertion* behind; and, hence, when it contracts, the foot becomes shortened, and the hinder part moves up towards the fore part. The *pallial* muscles consist of fibres inlaid in the edges of the mantle.

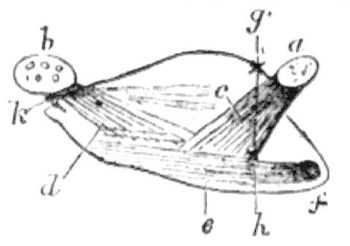

Fig. 80. — Showing disposition of muscles and ganglia in anodon. *a*, Anterior, *b*, posterior, adductor; *c*, anterior, *d*, posterior, retractor; *e*, protractor of foot; *f*, insertion of protractor of foot; *g*, *h*, *k*, cerebral, pedal, and parieto-splanchnic ganglia, connected by commissural cords.

The muscular fibres are composed of spindle-shaped bands (fig. 81), each constituting a single cell, with a nucleus. The protoplasm is filled in with granules, generally presenting the appearance of dots arranged in regular lines. These dots are *sarcous elements*, affecting polarized light in the same way as the cells of *striped muscle* in the higher animals. There is thus observed in the anodon a differentiation into sarcous elements and indifferent tissue.

* The *origin* of a muscle is the end at which it acts as from a fixed point; the *insertion* is at the opposite end, which, in contraction of the muscle, is drawn up nearer to the fixed end.

110 GENERAL BIOLOGY.

In the regularly disposed dotted arrangement, there is a rudiment of striped muscular fibre. In *form* these bands resemble smooth muscle, but in *structure* they belong rather to the striped variety.

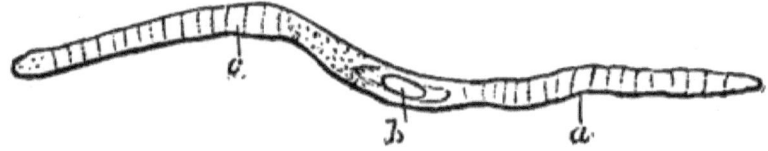

Fig. 81.—BAND-LIKE CELL OF MUSCULAR TISSUE TAKEN FROM ANODON. *a*, Cell; *b*, nucleus; *c*, "sarcous" elements.

168. Nervous System.—The anodon has three pairs of nervous ganglia—two *cerebral*, two *pedal*, and two *parieto-splanchnic*. The cerebral ganglia lie close to the anterior adductor muscle at either side of the œsophagus; the pedal ganglia are situated in the foot; and the parieto-splanchnic close to the posterior adductor. The œsophagus passes between the two nervous cords which unite the cerebral ganglia. Similar cords (*commissures*) pass from these ganglia to the pedal and parieto-splanchnic at either side.

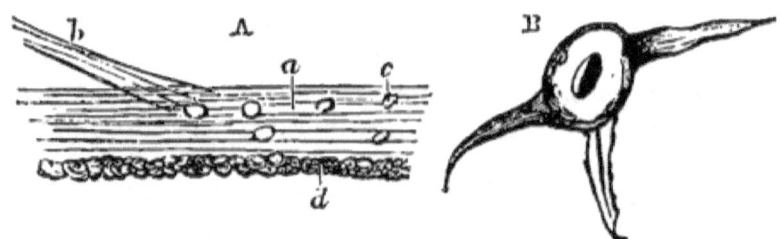

Fig. 82.—A, Nerve fibre in anodon: *a*, fibrillæ, seen branching at *b*; *c*, granules; *d*, indifferent tissue. B, Ganglionic corpuscle.

169. A nerve fibre, when examined under the microscope, is observed to be composed of exceedingly fine fibrillæ (fig. 82, A). These fibrillæ divide and form branches, which go to constitute a new nervous fibre. The nerves in anodon resemble very much in structure the sympathetic and olfactory nerves in the higher animals.

Minute granules A, c, are also seen dispersed through nerve fibre. At particular places, as in the ganglia, ganglionic corpuscles are found like that represented (fig. 82, B). By their means stimuli exerted on the body of the animal produce motion. The only organ of special sense found in anodon is that of hearing. It consists of a sac containing fluid and nerve matter, and a hard substance moving about in it called an otolith (fig. 83). This otolith, by impinging rythmically upon the nerve matter, produces the sensation of sound.

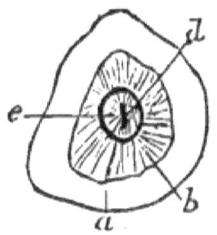

Fig. 83.—OTOLITH IN ANODON. *a*, Auditory capsule; *b*, cilia; *d*, otolith; *c*, fluid.

170. Reproduction.—The anodon breeds in winter or early spring. The reproductive organs are very simple, and their general character is the same in both sexes (for the anodon is bisexual). Close to the organ of Bojanus there is an aperture, into which lead a number of tubuli, lined with epithelium. These tubuli branch out into the substance of the foot, and at the breeding season (January or February), the cells which they produce are so well developed that the body becomes distended with them. Those in the male become flattened and take cilia, and pass into the epibranchial chamber, whence they are carried by the currents outwards, through the cloacal or "exhalant aperture," in streams into the water. Subsequently they find their way, aided by the water current, into the pallial chamber of the female, and fecundate the ova contained in the ovary.

171. In the female, the ova are developed in the same way as the sperm corpuscles in the male. Each ovum takes a coat, out of which there is an open projecting tube. Into this tube the ciliated sperm corpuscle enters, and becomes mingled with the substance of the ovum. The ova, after fecundation, pass out by the aperture already referred to; but, instead of following the general course of the fluid, they drop into the *external* gills, filling up their compartments. Thousands of them are lodged in a single pocket. In the gills they are protected from external injuries, and aërated as well, while they are undergoing further development. If an ovum be examined soon after impregnation, it will be found that after the process of yelk division, there will appear (fig. 84, A) two triangular pieces, like valves, held together by a muscle. If this organism, at a somewhat later period, be removed from the ovum, it will present the appearance set forth in fig. 84, B. The valves are observed always opening and closing again. The two valves are often opened quite straight. These are, however, the only vital movements observable. If we take a sectional view of the open valves, we find that they are joined by a highly elastic ligament; and that the elasticity of the latter is counteracted by a series of transverse fibres passing over from one valve to the other. Near the hinge there appear ciliated cells. Two projections from the living matter inside form the rudiments of the pallium. A small papilla projects from the under surface, which eventually becomes the foot. The valves differ from those of the adult animal in having at their edges re-curved hooks, armed with spines. From the papilla there hangs a few filaments, which constitute the **byssus**, by which the young animal is enabled to *anchor* itself whenever it chooses.

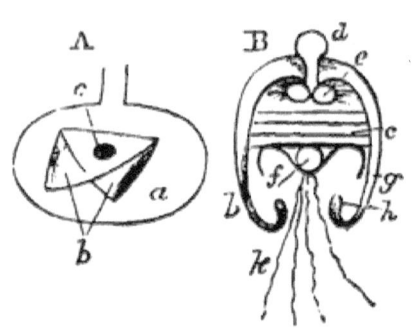

Fig. 84.—DEVELOPMENT OF ANODON. A, Earliest stage; *a*, ovum; *b*, valves; *c*, muscle. B, A stage more advanced, while still within the egg; *b*, valve; *c*, transverse fibres; *d*, ligament; *e*, ciliated cells; *f*, papilla, rudiment of foot; *g*, rudiment of pallium; *h*, re-curved hooks of valves; *k*, byssus.

172. At the end of a certain period, the vitelline membrane bursts, and the young bivalves all come out and

form clusters in the gills, hanging to each other by their hooks. When they were first discovered thus crowding in the gills of the parent, they were naturally supposed to be parasites, called *Glochidia*. After leaving the gill cavities in the current of water which passes through them, they attach themselves to floating bodies, more especially to the tails of small fishes, by means of their silken filaments and barbed hooks. It is thus they become dispersed over large areas. The papilla, already spoken of, grows into a foot; gills also become developed, and the hooks disappear; either they become dissolved, or else they get covered over with a new layer of shelly matter as the animal continues to grow.

173. The development of the oyster is somewhat different. After the process of yelk division, a cell is produced provided with a tentacle and cilia, which enable it to move about for some time. Then valves and a foot become developed. At last, numbers of these young oysters come forth "in little masses like drops of grease formed of several united together by an adhesive fluid," and attach themselves to rocks or adult shells. In this early condition they are called *spat* by the fishermen.

CHAPTER XIV.

THE LOBSTER (AN ARTHROPOD).

174. The Lobster (*homarus vulgaris*) in its structure and physiology is typical of the *arthropod* division of the sub-kingdom *Annulosa*. The arthropoda comprise the four classes, *myriapoda*, *insecta*, *arachnida*, and *crustacea*, all characterised by the possession of jointed body appendages. To the latter class the lobster belongs.

175. Morphology.—It is observed that this animal also is provided with an exoskeleton, but one very different, indeed, from what we have observed in anodon. The "crust" or hard part in the lobster contains twenty segments. Hence the body may be regarded as consisting of **twenty somites**, each "somite" being formed of a body part, with its appendages. Of these somites, six form the abdominal region (fig. 85), with the tail or telson *e*, which is not regarded as one of the body segments. They differ from the somites of the rest of the body in being articulated to, and hence movable upon, each other, by means of a flexible portion of the external integument *h*, and also in the form of the appendages. If we take one of the abdominal rings and examine it, we shall observe that its transverse section will be somewhat like the accompanying figure (fig. 86). The arterial blood system is observed to be near the dorsal side of the animal, the ganglionic cords or nerves are near the opposite side, while the viscera lie in the cavity between. The upper part of the circumference of the ring is called the **tergum**, the under part the **sternum**, and the part where both unite at a sharp angle is termed the **pleuron**. Each segment is provided with a pair of appendages, planted in sockets *g*, *h*, *k*, the basal part of each, *g*, being termed a **basipodite** or **protopodite**. In four out of the six segments which make up the abdominal part of the animal, namely, the second, third, fourth, and fifth, there are attached to each basipodite two little flappers *h*, *k*, like oars, having beard-like filaments hanging from their edges. The outer of

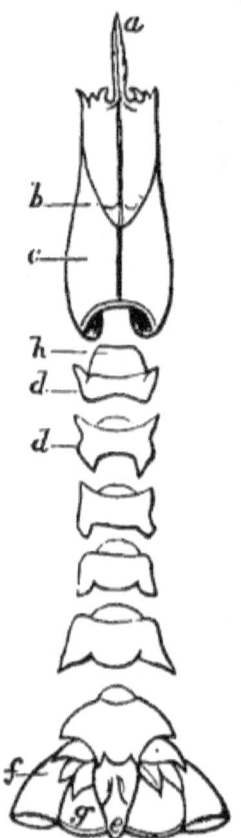

Fig. 85.—BODY SKELETON OF LOBSTER, with the segments detached and set in order. *a*, Rostrum or beak; *b*, cervical suture; *c*, carapace; *d*, *d*, abdominal rings; *e*, telson; *f*, *g*, "flappers;" *h*, flexible part of abdominal body ring.

MORPHOLOGY OF LOBSTER.

these, h, is called the **exopodite**; the inner, k, **endopodite**. In the first segment of the abdomen there is only one part to represent the exopodite or endopodite. In the sixth or last segment the basipodite is short and wide, and supports two large flappers (fig. 85, f, g), analogous to the exopodite and endopodite, one of which f—the exopodite— is again divided into two parts by a kind of hinge.

The movable rings overlap each other from front to rear, and are attached to each other by ligaments, and to the body by strong muscles.

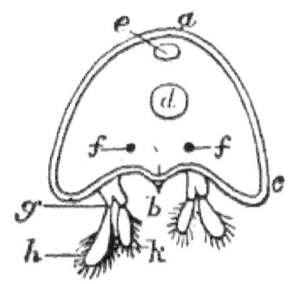

Fig. 86.— DIAGRAMMATIC SECTION OF AN ABDOMINAL RING OF A LOBSTER. a, Tergum; b, sternum; c, pleuron; d, alimentary canal; e, dorsal artery; f, f, ganglia; g, basipodite or protopodite; h, exopodite; k, endopodite; g, h, k, form a swimmeret.

176. If we examine the fore part of the exoskeleton, we find it to be made up of one piece, called the **carapace** (fig. 85, b, c), enveloping both the head and thorax. Hence this part of the body is called the **cephalo-thorax**. The cephalo-thoracic region is made up of fourteen somites, eight of these being distributed to the thorax, and six to the head. It is only on the sternal surface that the division of the cephalo-thoracic region into somites is well marked. On the tergal or dorsal surface the boundaries are obliterated, except at what is called the **cervical suture** (fig. 85, b), which forms the dividing line between the head and the thorax. The somites of the thorax are remarkable for the pleura becoming so very largely developed as to form between themselves and the body what is called the branchial cavity. Each pleuron thus officiates as a covering or protection to the gills, and is hence termed the **branchiostegite**. The general outline of the sternum, when the animal is extended at full length, is a horizontal line as far as the mouth. But in front of this the sternal line is turned upwards at right angles, causing those appendages which are situated in front of the mouth to be directed forward. This turn in the sternal line is called the *cephalic flexure*—a distinguishing mark in all the arthropoda. The head terminates in a beak projecting over the eyes, and termed the **rostrum**. The thoracic region bears eight pairs of appendages—five pairs of **ambulatory limbs**, and three pairs of **maxillipedes** or *foot-jaws*. The latter are so named because they subserve to the function of manducation.

177. Attached to each of the twenty somites is a pair of appendages. They are formed upon a common type, but modified so far as to be adapted to different conditions, and often to serve very different purposes.

178. General enumeration of the somites and their appendages :—

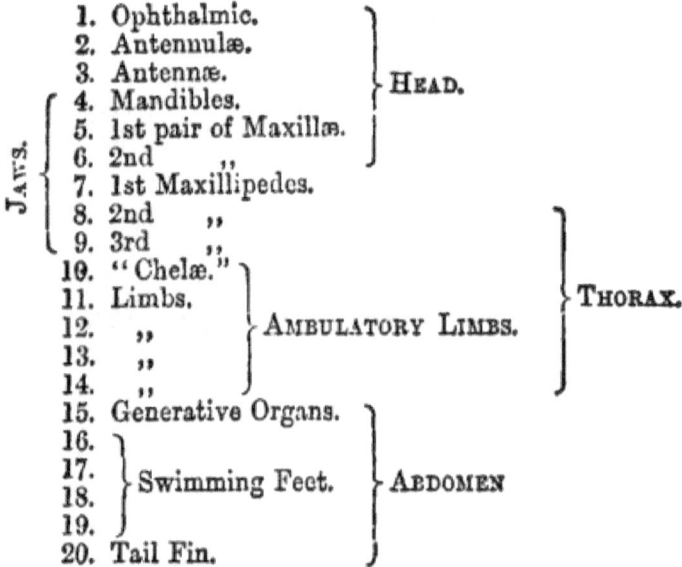

179. As a starting point, we may take the second maxillipede (fig. 87, A) as the type of the appendages, and compare the others with it, noticing in what they agree with, and in what they differ from, each other.

180. We observe that from a basal part *a* (the protopodite), there arise three structures like limbs, called, as

they are in the figure, *b*, endopodite; *c*, exopodite; *d*, epipodite. Now, if we examine the other appendages, we shall find that though different in appearance, they are all formed upon the above type, modified in such a way as to have some of those parts enlarged, reduced, altered in shape, or entirely suppressed.

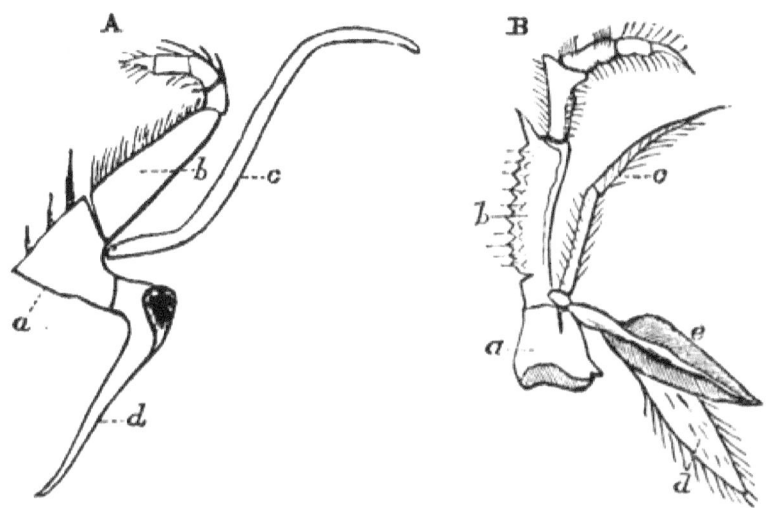

Fig. 87.—A, Second maxillipede, one of the second pair of foot-jaws: *a*, basipodite or protopodite; *b*, endopodite; *c*, exopodite; *d*, epipodite. B, Third maxillipede: the epipodite carries a gill, *e*.

In the third pair of maxillipedes (fig. 87, B), we find all the three structural parts present, the endopodite being serrated, for tearing food; but in the first pair of ambulatory limbs—the chelæ (fig. 89, I)—one of the parts becomes suppressed, namely, the exopodite. Notwithstanding this, however, we find that in young lobsters the exopodite takes its usual place. We have here, then, an instance of the suppression of parts. The exopodite is also wanting in the remaining four ambulatory limbs.

The second and third pairs of ambulatory limbs (fig. 89, II, III), are also chelate, but they do not attain the dimensions of the "chelæ" proper, more especially in the "pincers."

If we now proceed to examine the structure of the fourth, or last but one, of the ambulatory limbs (fig. 89, IV), we shall see how the pincers in the chelate limbs become developed. If we suppose the little projection at the last joint to extend outwards, and to place itself in apposition to the claw, we shall have the "pincers." We have thus in the "pincers" of the chelate limbs an example of a morphological change, by the greater development of a particular organ. The epipodite is wanting in the next limb—the eighth thoracic—which is also non-chelate.

In the first abdominal appendage, all the parts are also suppressed except the endopodite, which is so far modified as to be used as part of the genital apparatus. The remaining abdominal appendages have been noticed already.

It then becomes manifest that all the observed differences in external appearance in the limbs are due to the modification, suppression, or enlargement of some part of the typical limb.

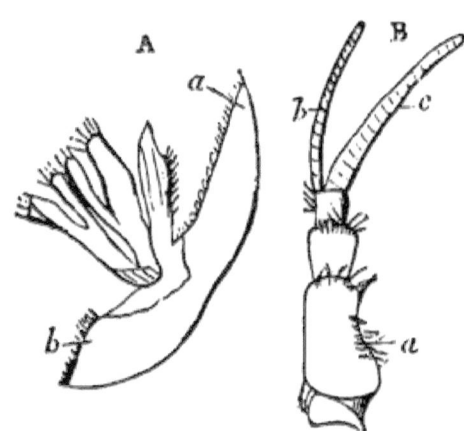

Fig. 88.—A, Second maxilla or scaphognathite; *a*, exopodite; *b*, endopodite. B, Antennule of lobster; *a*, basipodite; *b*, exopodite; *c*, endopodite. These latter parts are largely cut away, only a small portion appearing in the figure.

181. The second maxilla or "scaphognathite" (fig. 88, A), is used to bale out the water of the branchial cavity on either side. This cavity communicates with the exterior by two apertures. Behind there is a cleft between the base of the limbs and the side of the gill cover (*branchiostegite*). The water enters here, passes over and among the gills, and is baled out at the extremity of a

groove in front, by the scaphognathite. In this way, a constant current of water is maintained through the gills, such as is produced by ciliary action in the anodon. In this appendage, the epipodite and exopodite become united to form the "baling" apparatus. It is presumed that the foliated part is simply formed of lateral processes of the basipodite.

Passing forward to the *first maxilla* (see fig. 103, B), we find a large basal part, supporting a comparatively long exopodite, and a foliated endopodite—the epipodite wanting. The next pair are the *mandibles*. Each consists of a comparatively large basal part, and a small segmented endopodite, which is called the *mandibular palp*. The mouth opens between the two mandibles. The next pair of appendages in front of the mouth are the *greater antennæ*, or "long feelers." They are sometimes longer than the animal's body; at their bases are observed the "green glands," the supposed renal organs of the lobster. Each antenna is formed of a segmented endopodite; the exopodite, which occurs in the shrimp, being in the lobster reduced to a mere scale. Next in forward order are the **antennules** or *lesser antennæ* (fig. 88, B). These contain at their base, the organs of hearing; they consist of a basal part with an exopodite and endopodite segmented, and of nearly equal length. They are not nearly so large as the antennæ. The eyes, with their peduncles, are the appendages of the first somite of the body. At the top they have the chitinous coating removed, and replaced by the transparent *cornea*, which is divided by cross lines into a number of squares or facets, each serving the purpose of a lens.

182. We now proceed to describe the arrangement of the organs within the body. By taking a cross section, say, through the heart, we observe that the heart and principal vessels are situated near the *tergal* side of the body. The alimentary canal follows the course of the main axis, and the nervous ganglia, with their commissures, run along near the sternal or under side.

183. In the cephalo-thoracic region there is a device for the protection of the nervous ganglia, which is not found in the abdomen. There exists a partition of hard matter, separating the ganglionic cords from the chamber over them, somewhat as the neural arches protect the spinal cord in the vertebrate animals. This **endophragmal partition**, as it is called, is formed by an ingrowth of the substance of the exoskeleton, meeting in the middle line, and forming a kind of *roof* for the ganglionic chamber. There are six pairs of ganglia in the abdominal segments, six in the thorax, and one in front of the mouth—thirteen in all.

184. By cutting off the back in the thoracic region, we observe a sac filled with fluid. This is the pericardium; inside of it is the heart; one artery goes backwards, along what may be called the dorsal side, another runs forwards to supply the eyes, and the anterior parts

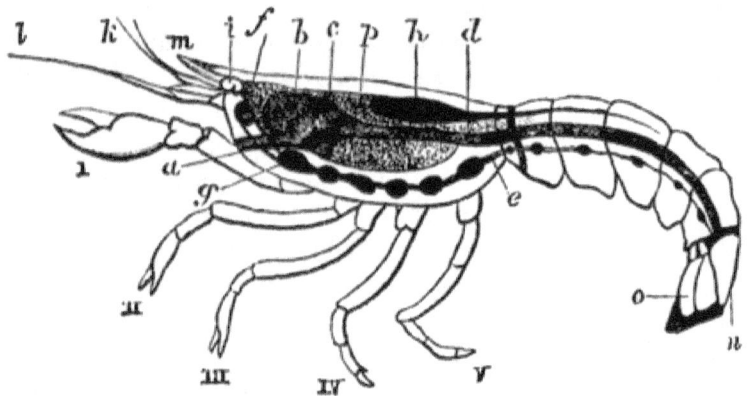

Fig. 89.—Vertical and Longitudinal Section of Lobster, showing the alimentary and nervous systems, and ambulatory limbs. *a*, Mouth; *b*, cardiac; *c*, pyloric cavities of stomach; *d*, intestinal canal; *e*, chain of ganglia; *f*, cerebral ganglion; *g*, hypœsophageal ganglion: the "oesophageal collar" is seen surrounding the gullet. The ganglia that succeed are those of the thoracic somites. Behind *e* are the ganglia of the abdominal somites. I, II, III, IV, V, are the ambulatory limbs; the first three chelate, the others not. *h*, Heart; *i*, eye; *k*, antennule; *l* antenna; *m*, rostrum; *n*, telson; *o*, flappers; *p*, liver.

of the body. From the posterior artery, a third turns downwards, becoming what is called the *sternal* artery. There are twenty gills on each side, partly attached to the basal joints of the thoracic appendages, partly to the side wall of the thorax.

185. **Physiology—Alimentation.**—The **mouth** (fig. 89, *a*), which is lined with chitin, is situated between the **labrum** in front, and the **metastoma** or **labium** behind, corresponding respectively to the upper and under lips of vertebrate animals. The chitinous lining extends along the stomach and intestines. The anterior or *cardiac* part *b* of the stomach contains within its cavity an apparatus which is ancillary to mastication. It consists of three teeth, which, by a very complex mechanism, meet each other in the middle line to crush and triturate the hard parts of the food. The posterior or *pyloric* orifice of the stomach will not allow any but the most comminuted food to pass through. There is a small triangular space at this point, which may be diminished to any extent by the pressure of the three bodies which surround it. To this is superadded a sort of sieve or strainer formed of hairs crossing each other, which will not allow any but the smallest particles to pass through them.

186. In all vertebrate animals, the muscles of the stomach and intestines are composed of smooth fibre, but in the lobster, and, indeed, in all the arthropoda, all the fibre is of the striated kind. The intestine is lined with an epithelial layer; next to this is a layer of chitin, then circular and longitudinal fibres, and outside, again, a fibro-vascular coat.

187. The **liver** (fig. 89, *p*) is a large yellowish mass; it is duplicate, one part being on each side of the middle line. There are two ducts opening from it into the alimentary canal. These ducts pour out the secretions of numerous cæcal tubuli (fig. 90, A, B), lined with granular cells. It is in these granular matters that the secreting power resides. There are no other glands connected with the alimentary canal, except one blind tube

opening into it near the anus. Its function has not yet been determined. The intestine in the lobster is quite straight, and terminates at the root of the telson.

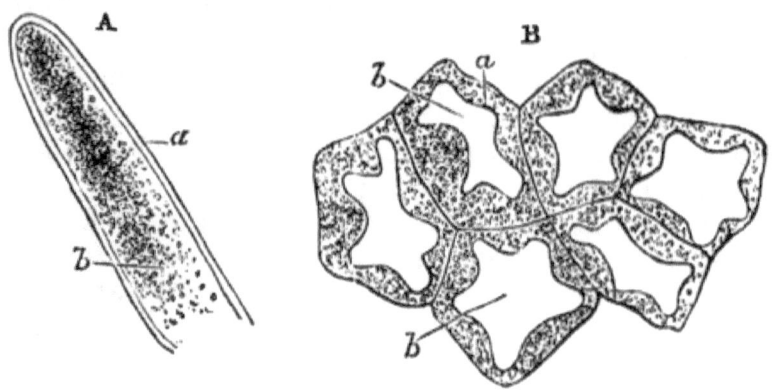

Fig. 90.—A, Single tubule of liver treated with chromic acid and glycerine; *a*, sac; *b*, granules. B, Cross section of six apposed tubuli; *a*, cellular structure; *b, b*, cavities receiving the secreted matters.

188. Circulation.—The heart, as has been already observed, lies in the dorsal part of the thorax, and is enclosed within a wide cavity called the pericardium (fig. 91, A, B). This chamber differs from the pericardium of vertebrates, principally in the fact that it is a *blood sinus*, and discharges also the functions of an auricle, communicating with the heart—which in the lobster is entirely ventricular—by no fewer than six openings. In vertebrates, the pericardium is a closed *sac*, containing *serum*, not blood, and having no direct communication with the chambers of the heart. The pericardium of the anodon also differs from that of the lobster, in the circumstance that the former is both a blood and water reservoir, and communicates directly with the exterior by the channels hollowed out in the organ of Bojanus. The heart in the lobster is connected with the inner wall of the pericardium by bands of muscular fibres *c*, which must not be confounded with blood-vessels. Muscular fibres also traverse the cavity of the heart itself, giving to its interior a spongy appearance. The heart communicates with the pericardium by three pairs of apertures, one pair on the back *d*, one at the sides *e*, and one on the under floor *f*. The edges of these apertures turn inward, to serve the purpose of valves. When the muscles connecting the pericardium with the heart *contract*, the latter becomes *dilated*, and blood rushes into it from the pericardium. The action of the muscles in the interior of the heart then causes it to *contract*, and drive the blood contained in it into the arteries. The pericardium contracts but little, if, indeed, it do so at all; still it must be regarded as discharging the functions of an auricle in acting as a reservoir for the blood previous to its reception into the heart.

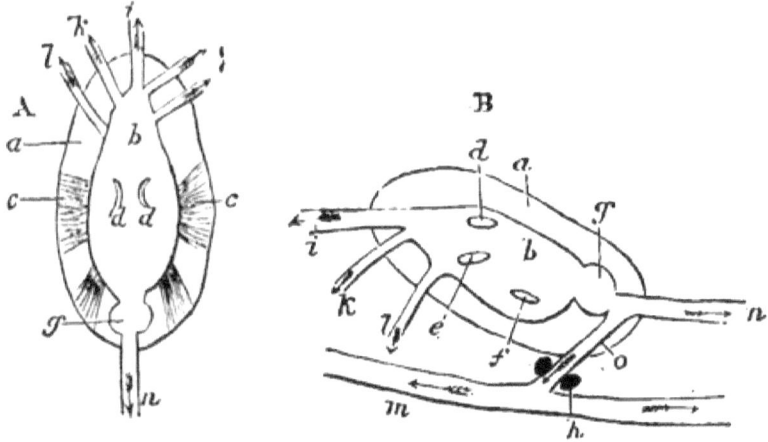

Fig. 91.—DIAGRAMMATIC VIEW OF THE HEART AND GREAT VESSELS IN A LOBSTER. A, As seen from above. B, Side view. The same letters indicate the same parts. *a*, Pericardium; *b*, heart; *c, c*, muscles uniting it with pericardium; *d, e, f*, superior, lateral, and inferior openings leading into heart; *g, bulbus arteriosus; h*, ganglion of 13th somite; *i, k, l*, arteries leading to eyes, antennæ, and liver (or stomach) respectively; *m*, sternal artery; *n*, superior abdominal artery; *o*, artery connecting these two.

189. The posterior aorta, as it leaves the heart, becomes dilated into a **bulbus arteriosus** (fig. 91, *g*), having strong muscular walls, which enable it to propel the blood through the two arteries which spring from it. One of these, the superior abdominal artery (fig. 92, *b*), passes backward along the median line, under the dorsal integument, giving a pair of branches to each somite. The other artery *c* connects the bulbus arteriosus with the **sternal artery** *d, d*, which it joins between the ganglia of the thirteenth somite (fig. 91, B, *h*). From this point the blood is propelled both forward and backward to supply the under parts of the body. Three arteries proceed from the *anterior* end of the heart—one along the median line, giving branches to the eyes (figs. 92, *e*, and 91, *i*), and one on each side (figs. 92, *f*, and 91, *k*), to supply the antennæ. A second pair leave the under floor of the heart to furnish blood to the stomach and liver (fig. 91, *l*).

190. The arteries do not terminate in capillaries as in vertebrate animals, but in irregular sinuses or "lacunæ," hollowed, as it were, out of the tissues. Thence the blood collects into larger sinuses, and passes at length into the great venous cavity—the **sternal sinus** (fig. 92, *h*)—from which it is conveyed to the gills at *g*.

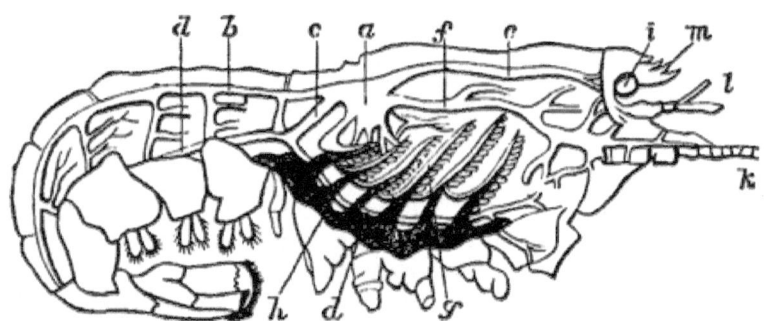

Fig. 92.—SIDE VIEW OF INTERIOR OF LOBSTER to show circulation. *a*, Heart; *b*, superior abdominal artery; *c*, artery leading to *d*, sternal artery; *e*, anterior median artery, supplying eyes; *f*, artery leading to *k*, one of the great antennæ; *g*, gills; *h*, sternal sinus; *l*, antennule; *m*, rostrum.

191. **Respiration.**—Each **gill** consists of a main stem, and little branches like a bottle brush. The stem is

hollow, and is divided into two distinct passages; each little branchlet is also hollow, and contains two separate canals. The venous blood which passes from the sternal sinus into the gills takes only one of the canals. It there becomes aërated, and returns to the base of the gills by the second canal, and then proceeds to the pericardium, whence it enters the heart. The heart in systole forces the blood into all the arteries, both backward and forward; but it receives no venous blood till after its oxygenation in the gills. The heart is thus entirely *systemic*, as is the case in all the arthropoda — not *branchial*, as in fishes. In the blood of arthropods, as in that of all invertebrate animals possessed of a blood circulation, corpuscles of only one kind are found, namely, the colourless.

192. The epipodites of the thoracic appendages lie among the gills, and some of the gills are attached to them. The more actively the animal moves his limbs, the more readily is his blood oxygenated, and his worn-out tissues wasted.

193. It is not yet settled as to where the waste *nitrogenous* elements are excreted from the blood; but it is supposed that this function is discharged by two green glandular masses (fig. 93), situated at the base of the great antennæ. When examined, each mass

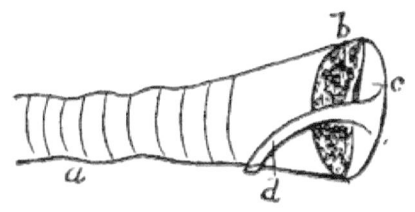

Fig. 93.—"GREEN GLAND" AT FOOT OF GREAT ANTENNA, supposed renal apparatus. *a*, Antenna; *b*, glandular structure; *c*, water reservoir; *d*, excretory duct.

consists of a cavity *c*, containing water, which cavity opens at the base of the antenna, together with a sac-like apparatus *b*, which is supposed to be the excretory organ.

194. **The Muscular System.**—In the lobster all the muscles are of the *striped* variety. Each muscular fibre (fig. 94) consists of a number of fibrillæ, enclosed in a

sheath (**sarcolemma**) of connective tissue. It will be observed that each fibrilla is divided by transverse bars into a number of segments; each segment seems again to be divided by a faint line. On examining a limb, we find its several joints to be articulated so as to move in different planes. This gives a much wider area of motion to the limbs than would be the case if all the joints allowed of motion only in one plane.

195. The joints are not constructed on the same plan as in vertebrate animals. They remain uncalcified at two points, and hence allow of a certain degree of flexibility at those parts.

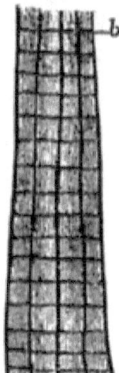

Fig. 94.—STRIATED MUSCLE IN LOBSTER.

196. The muscles do not act on a limb in the same way as in vertebrate animals. Suppose the upper portion of the limb in fig. 95 to be movable upon the under one. When the muscle marked A contracts, the direction of the movement in the upper limb will be towards a. On the other hand, the muscle marked B would move the limb in the direction b. This is directly *opposite* to the direction it would take were the muscles external to the limbs, as they are in vertebrate animals, instead of being placed inside, as in the arthropoda. All the movements of the muscles are directed by the nervous system.

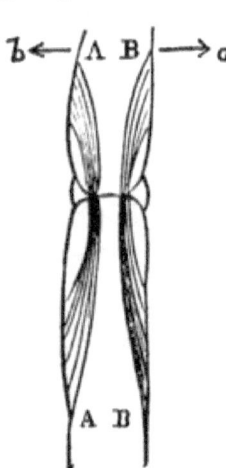

Fig. 95.—SHOWING HOW A MUSCLE FLEXES A JOINT.

197. **Nervous System.**—By referring to fig. 89, it will be observed that in the lobster there is a double chain of ganglia running along the *under* side of the body, only one of which is above the alimentary canal, namely, the cerebral. As there are twenty somites, we should ex-

NERVOUS SYSTEM OF LOBSTER.   127

pect to find twenty pairs of ganglia; but those of the first nine somites are collected into two large masses; one just over the mouth, called the cerebral ganglia (fig. 96, *a*), and one under the œsophagus and behind the mouth, called the hypœsophageal, *b* (see also fig. 89, *f, g*). These two masses are connected by two nervous cords (fig. 96, *r, r*), one on each side of the œsophagus. The latter organ is thus surrounded with a ring of nervous matter, which is hence called the **œsophageal collar.** The remainder of the thoracic region has five pairs of ganglia, one to each somite. Each of the abdominal segments is also provided with a pair. There are thus thirteen pairs in the chain, two of them being masses formed by the coalescence of nine pairs of original ganglia, making twenty pairs in all. In the crab, coalescence goes much farther than this, all the post-œsophageal ganglia uniting into one great mass of nerve matter. This may be accounted for by the diminished length of the body.

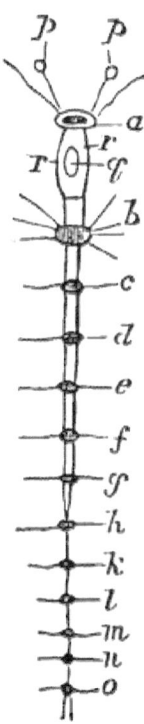

Fig. 96. — NERVOUS SYSTEM IN LOBSTER. *a*, Cerebral ganglia; *b*, post-œsophageal ganglia; *c, d, e, f, g*, thoracic ganglia; *h, k, l, m, n, o*, abdominal ganglia; *p, p*, eyes; *q*, passage for œsophagus; *r, r*, commissures forming the "collar."

198. In the lobster there are two sorts of nervous matter—**nerve cells** and **nerve fibres** (fig. 97, A, B). Some of the latter are much larger than any to be found in the higher animals, amounting to about $\frac{1}{200}$ of an inch in diameter; but many of them are much smaller. When a smaller nerve fibre is examined under the microscope, it is found to consist of a very thin wall, enclosing a clear, transparent glass-like substance, dotted over with minute granules, which keep moving through the mass, thus proving it to be completely fluid. The fibre is also observed to contain numerous corpuscles or *nuclei*, which usually attach themselves to the wall or sheath. The larger nerve fibres, however, have thicker walls. In both classes of fibres, the wall is entirely structureless. A bundle of nerve fibres is usually enclosed in a sheath, called a **neurilemma**. The cells of the nervous ganglia are nucleated, and give off processes which pass into nerve fibre.

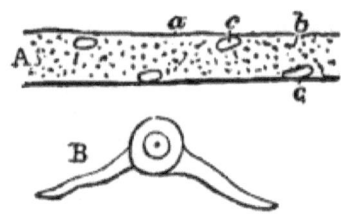

Fig. 97.—A, Nerve fibre : *a*, structureless membrane ; *b*, granules; *c, c*, corpuscles or "nuclei." B, A nerve cell.

199. Some of the nerves are distributed to the vessels of the body, and to the muscles; some to special organs of sensation; but there is no grouping, as in vertebrates, into afferent and efferent bundles. Of the sensory nerves, some pass into the ophthalmic appendages, to receive the impressions of light by means of the eyes; some into the base of the antennules, to render the organ of hearing sensitive to sound.

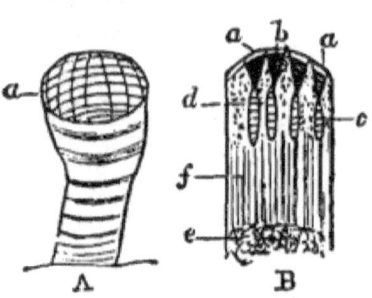

Fig. 98.—A, Eye of lobster, showing *a*, cornea, with its cross markings. B. Vertical section of same, showing *a*, facets; *b*, cones inverted; *c*, rods, exhibiting cross markings; *d*, pigment granules ; *e*, nerve matter ; *f*, nerve filaments distributed among the rods and cones.

200. The lobster is **podophthalmate**, that is, his eyes are supported on movable "peduncles" or eye-stalks, which may be turned in various directions at the will of the animal. The cornea (fig. 98, A, *a*) is divided by cross lines into a number of little squares or "facets," each of which is a separate sensory apparatus,

supposed even by some to be a separate eye. From each "facet" there hangs an inverted cone B, *b*, to the apex of which there is a prismatic rod-like body B, *c*, of a gelatinous nature attached, which shows several cross markings. The interspaces are filled up with pigment B, *d*, and connective tissue, among which minute nerve filaments B, *f*, are distributed, as seen in the figure. From the general resemblance these rods and cones bear to those in the eyes of vertebrate animals, it is supposed that one of these compound eyes, taken as a whole, has the same effect as, for instance, the human eye, whose retina contains a multitude of such rods and cones.

201. The **auricular apparatus** or "ears" are placed at the base of the antennules; each consists of a bag (fig. 99, *a*), holding a clear liquid, with some grains of sand, "otoconia." It communicates with the exterior by a small cleft in the antennule. The interior of the sac is covered with minute filaments, and abundantly supplied with nerve. When a sound is conveyed, the vibrations are communicated to the fluid, which thus causes the sand-granules to move about rythmically through it. In doing so they impinge against the sides of the sac, or rather against the filaments disposed along its surface, and in this way convey to the nerves the impulses created by the vibrations. The "otoconia" are not secreted by the animal; they are simply grains of sand received into the auditory cavities opening on the antennules, and may be increased or diminished in amount as circumstances may require.

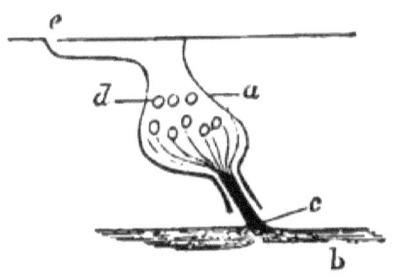

Fig. 99.—AUDITORY APPARATUS OF LOBSTER. *a*, Auditory sac; *b*, nerve, giving a filament *c* to the auditory capsule; *d*, otoconia; *e*, external meatus, being a cleft in antennule.

202. Reproduction.—The organs of reproduction are larger in the female than in the male. In the latter, the *spermarium*, or gland which secretes the sperm corpuscles, is a long tube-like body, of which there are two, one on either side. The aperture of exit for the sperm cells is at the base of the fifth pair of ambulatory limbs. These cells are not furnished with vibratile cilia, as in fern and anodon. Each consists of a spheroidal body, with three filaments projecting from one extremity; but it possesses no power of locomotion. This, indeed, is the case with the sperm corpuscles in all the arthropoda.

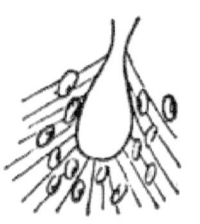

Fig. 100.—Showing ova clustering on the swimmerets of the female lobster.

203. In the female the ovary is a dark green mass, which, at the breeding season, fills a very large part of the body. The duct is at the base of the third pair of limbs. At the time of laying she bends up her abdomen under her body; the eggs thus become attached (fig. 100) to a viscid substance, which covers the hairs of the abdominal appendages. They are thus carried about with the female, and duly aërated until they become hatched.

Fig. 101.—Embryo Lobster (*Zoea*) as it leaves the egg.

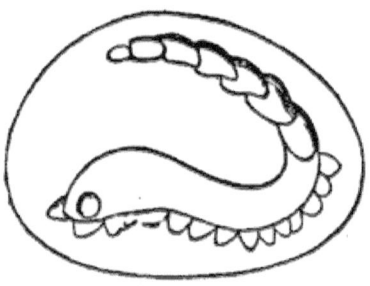

Fig. 102.—Embryo Cray-Fish in the Egg. It is observed that the ventral side is directed outwards, and that the umbilicus is at the dorsal side—the reverse of what is observed in vertebrate embryos.

REPRODUCTION OF LOBSTER. 131

After the usual breaking up of the matter of the ovum, by endogenous cell multiplication, there is developed within it an embryo lobster, somewhat resembling the first representation set forth in fig. 101. There are no abdominal limbs. In this larval condition, in which it gets the name of *zoea*, it is set free, after which it casts off its skeleton, and secretes a larger one. These operations are repeated time after time, while the zoea is

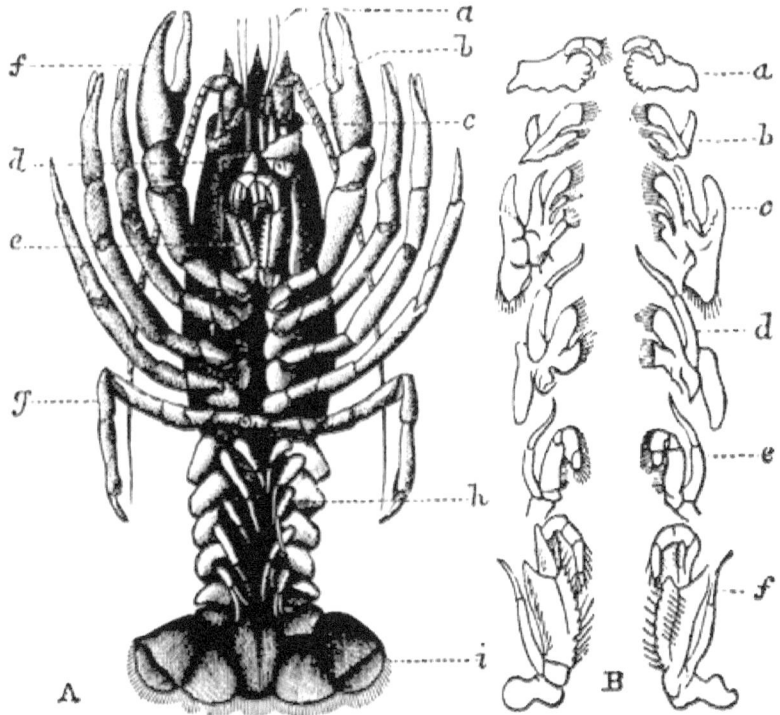

Fig. 103.—A exhibits the lower side of the Cray-fish : *a*, antennule; *b*, antenna; *c*, eyes; *d*, auditory tubercle or organ of hearing; *e*, external foot-jaws ; *f*, first pair of thoracic legs ; *g*, fifth pair; *h*, abdominal false legs; *i*, tail formed for swimming. B shows, in their detached state, the six pair of appendages which constitute the apparatus for mastication: *a*, mandibles; *b* and *c*, first and second pair of jaws or maxillæ; *d*, *e*, *f*, three pair of foot-jaws, gradually approaching the form of the ordinary limbs.

growing, and undergoing successive metamorphoses, until at last it assumes the form and proportions of the adult animal.

204. In the **cray-fish** (*astacus fluviatilis*), of which a representation is given in fig. 103, the mode of development is somewhat different. The yelk breaks up only at or near the surface. Twenty cellular bud-like processes (fig. 102) appear on the *neural*, that is, the ventral side of the embryo, which in invertebrate animals is always turned outwards, never towards the centre of the ovum. The umbilicus is therefore situated posteriorly. As the embryo progresses in growth, the abdominal joints become extended, assume processes or appendages, the thoracic and cephalic members soon after show themselves, and thus development proceeds until the form of the adult animal is ultimately attained, before even the embryo has escaped from the ovum. Hence there is no metamorphosis in the development of the cray-fish.

As the adult cray-fish is similar in organization to the lobster, little need be said of it except that it differs from the latter in being an inhabitant of fresh water, generally living in hollows excavated in river banks, unless when in pursuit of prey. This consists of molluscs, aquatic insects, etc. Fig. 103 is a representation of one, viewed from below. The figure also gives a separate view of the mandibles, maxillæ, and foot-jaws, arranged in their natural order, from before backwards. It is to be remembered that the mouth is situated between the bases of the mandibles.

CHAPTER XV.

THE FROG (AN AMPHIBIAN).

205. IN the anodon we have had an example of a mollusc; in the lobster, of an arthropod; we next proceed to consider the structure and physiology of the **vertebrate** sub-kingdom, taking for our example the **frog**. This is not indeed the best *type* of a vertebrate animal, but it is selected because, *firstly*, it is easily procured and examined; *secondly*, at different periods of its life it exhibits the characters both of a fish and an air-breathing animal; *thirdly*, its mode of reproduction presents the fewest complications; and *lastly*, experiments made upon this animal have been the means of establishing many important principles in physiology. The frog belongs to the class **amphibia**, and to the order **batrachia** —also called **anura** (or *tailless* animals).

206. **Morphology.**—We notice that, like the anodon and lobster, the frog is bilaterally symmetrical; but we observe no obvious segmentation, such as that exhibited by the lobster. In considering its parts we may regard it as being made up of a **trunk** and two pairs of **limbs**. The trunk again may be divided into **head, neck, thorax,** and **abdomen**. A number of apertures are likewise observable, some arranged along the median line, and hence single; others laterally and in pairs, one on each side. Besides the ordinary eyelids, the frog has a third eyelid, which in man is only rudimentary, the *nictitating* membrane. This is a strong fold, fixed at the anterior side of the eye, and capable of being rapidly drawn over the eyeball, beneath the ordinary eyelids. The *tympanum*, or cavity of the ear, communicates freely with the throat, but is covered externally by a membrane (the tympanic membrane). The nasal passages open into the mouth, and are provided with valves, which may be closed at the will of the animal.

207. If we examine the *limbs* of the frog, we shall find that the fore and hinder pairs have a remarkable correspondence with each other. Placed in order, they may thus be compared one with the other:—

(*Fore-limb.*)	(*Hinder-limb.*)
Brachium (arm).	Femur (thigh).
Antebrachium (fore-arm).	Crus (shank).
Manus (hand).	Pes (foot).
Digiti (fingers or toes, 4).	Digiti (toes, 5).

208. The external integument of the frog is a soft and loose coat or **skin**. In the lobster this is the hard part of the animal; but in the frog, the hard part or *skeleton* is *inside*, and termed, for this reason, an **endoskeleton**. We observe also that the integument of the frog is quite naked, uncovered either with *scales, feathers,* or *hairs,* and kept constantly moist by numerous glands opening upon its surface. The claws or "nails" are developed like the nails of the human finger, by the agglomeration of epidermic scales. There are four on each fore limb, and five on each of the hinder, corresponding in each case to the number of digits.

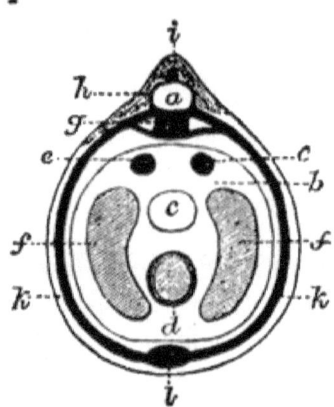

Fig. 104. — DIAGRAMMATIC SECTION OF VERTEBRATE ANIMAL THROUGH THE THORAX. *a,* Neural cavity, *b,* visceral cavity; *c,* alimentary canal; *d,* heart; *e, e,* sympathetic ganglia; *f, f,* lungs; *g,* centrum or body of vertebra; *h,* neural arch; *i, i,* spine; *k, k,* ribs; *l,* sternum.

209. Internal Structure.—If we make a transverse section through the thorax of a typical vertebrate animal (not, however, of a frog, for it has no *ribs*), we shall have such an appearance as is represented in the accompanying figure (fig. 104). We observe, in the first place, that the section will exhibit two cavities, one much larger than the other. The smaller of the two, *a,* is the **neural cavity**. It contains a long cord of nervous matter, passing the whole way from the brain to the posterior end of the body. This cavity is completely shut out from the larger cavity by a chain of bones, consisting of the **centra** or "bodies" of the vertebræ *g*. The larger cavity of the two, *b*, is called the **visceral cavity**. It contains the *viscera*, including the heart *d*, the lungs *f, f*, and the alimentary canal *c*. There are also two chains, *e, e*, of ganglionic nerve matter, running along the upper or "back" wall of the *visceral* cavity, on either side of the median line. These constitute the **sympathetic nerve system**. The section also gives a view of the heart and lungs, or vessels proper to the thorax. The thorax, in most vertebrates—not, however, in the frog—has its walls strengthened by the **ribs** *k, k*, which extend from the vertebral column above to the **sternum** *l* in front. That portion of the vertebral column which encloses the spinal cord gets the name of the **vertebral** or **neural** *arches*, *h*, while the more solid part of the column, beneath the spinal cord, is formed of the *centra*, or "bodies of the vertebræ." The whole body is covered with an **epidermic** layer, and the cavities are lined with **epithelium**.

210. It is observed that the heart and nervous system are on *opposite* sides of the body, having the alimentary canal between them. The same arrangement obtains in the lobster, and, indeed, in all the invertebrata possessed of a heart and nervous system. But there is this difference between the frog and the lobster in respect to this arrangement: in the lobster, the locomotive limbs are turned *towards* the same side as the nervous chain, whereas in the frog it is just the reverse. The same difference is observed between all the vertebrata, on the one hand, and all invertebrata endowed with a nervous system, on the other. If, in all classes of animals, we called the side on which the nerves are placed, the "back," then we should describe a lobster, a cricket, a spider, or worm, as *an animal that walked with its back downwards.*

Similar sections through other parts of the body would present a similar arrangement of parts, but the heart and lungs would, of course, disappear. In mammals, the thoracic region is separated from the abdominal by the *diaphragm*, which is a large muscular sheet attached to the vertebral column behind, to the sternum in front, and to the ribs on either side. As the frog, however, is not a mammal, it possesses no diaphragm.

211. The *pericardium* in the frog is like "a double nightcap" investing the heart with a double coat, the interior fold of which is attached to the heart at all points, and to the great vessels at their point of junction with it. The space between the two folds is filled with *serum*, which renders the movements of the heart smoother and easier. In the frog, the pericardium differs from that in the anodon and lobster in being completely closed, and in containing no blood within its own cavity. The *pleura* is another serous membrane, investing the lungs in the same way as the pericardium does the heart, but, in this instance, the outer coat serves to line the walls of the thorax.

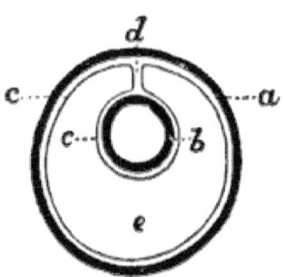

Fig. 105.— DIAGRAM SHOWING ARRANGEMENT OF PERITONEUM. *a*, Body wall; *b*, wall of intestine; *c, c*, peritoneum investing intestine and lining body wall ; *d*, mesentery supporting intestine; *e*, peritoneal cavity.

212. The alimentary canal in the adult frog is very simple and short, having but few convolutions. The *peritoneum* invests the intestines in a manner perfectly similar to the folding of the pericardium on the heart, and the pleura on the lungs. The canal is, in fact, supported by the folds of the peritoneum, as is seen in the annexed figure (fig. 105). The supporting part of the peritoneum is termed the **mesentery**. The cavity between its two folds is called the peritoneal cavity.

213. **Histology of Endoskeleton.**—The skeleton in

vertebrate animals may be made up in whole or in part of *bone, gristle* (or *cartilage*), or *connective tissue.* **Connective tissue** forms the general framework of the whole body, from the epidermis, on the outside, to the epithelium within. It has been said that if all the other materials of the body were removed, a perfect model of it would, nevertheless, remain, formed of connective tissue. In its fully developed condition, this structure is formed of two elements—two sorts of fibrous tissue, one elastic, the other non-elastic. In ordinary circumstances these two elements are undistinguishable, and when boiled the tissue swells up and yields **gelatine**. Acetic acid causes it to assume a similar appearance; but the elastic fibres are now perfectly distinct from the gelatinous part, as they are not acted upon by the acid. Cells or "nuclei" are also observable, and for the same reason. The fibrous part of connective tissue bears the same relation to the nuclei or cellular element, that the cellulose in plants, and the *lignine* formed from it, bear to the primordial utricle, inasmuch as both are differentiated materials, produced by the action, and at the expense, of the respective protoplasms.

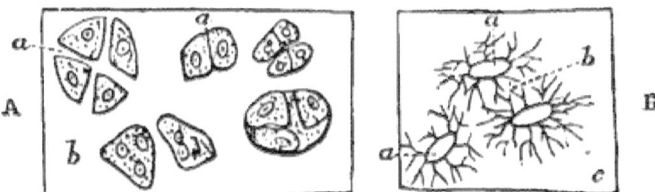

Fig. 106.—A, Cartilage under the microscope, showing nucleated cells, surrounded by a *matrix:* a, a, cells multiplying by fission; b, matrix formed from cells. B, Bone under the microscope, showing a, lacunæ; b, canaliculi; c, matrix.

214. Cartilage or **gristle** is quite a different material from connective tissue. It is translucent and elastic, and exhibits no evidence of a fibrous structure. When examined under the miscroscope it is found, however, not to be quite homogeneous, but to consist of a *matrix* (fig. 106, A), in which numerous nucleated cells are embedded. The matrix is manufactured by the cells, and at the expense of their protoplasm. It produces, on being boiled, a substance called **chondrine**, which is quite different from gelatine. There is, however, every form of transition between cartilage and connective tissue.

215. Bone is formed of a matrix (fig. 106, B), which gives either gelatine or chondrine on being boiled, but which is opaque from its being impregnated with earthy salts, principally phosphate and carbonate of lime. In the matrix there are cavities which contain air, when the bone is in the dry state; but when it is fresh these *lacunæ*, as they are called, are occupied by nucleated corpuscles. Bone is a secondary formation, produced by the deposition of calcium salts in *connective tissue*, or in *cartilage*.

216. If we regard their origin, then, there are two kinds of bone:—

1. Such as is modelled on a pre-existing connective tissue.

2. Such as has been pre-modelled in cartilage.

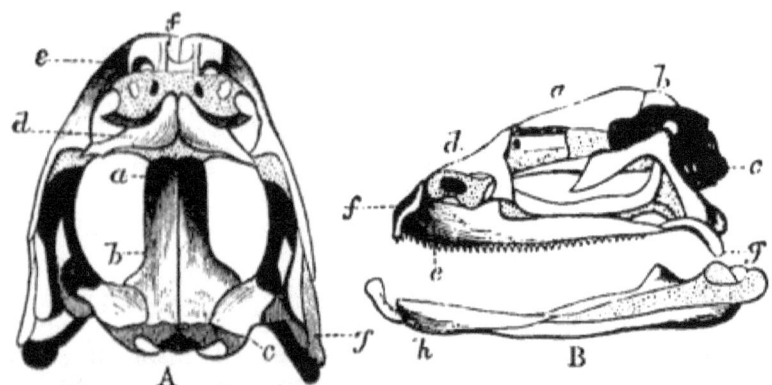

Fig. 107.—Skull of Frog. A, as seen from above; B, side view. *a*, frontal bone; *b*, parietal; *c*, exoccipital; *d*, nasal; *e*, maxilla or upper jaw; *f*, praemaxilla; *g*, quadrato-jugal bone; *h*, mandible.

Most of the bones of the skull have had their origin in

connective tissue, those of the limbs are formed by the deposition of salts of lime in *cartilage*. We may designate the former sort of bone, "membrane bones," that of the latter, "cartilage bones." The clavicle is usually a membrane bone.

217. The skeleton, as a whole, may be divided into two great divisions—

A. Axial skeleton, consisting of head and trunk.

B. Appendicular skeleton—the part formed by the limbs.

Our description of the skull, or cranial part of the skeleton, will go no further than merely to submit the illustrations A and B (fig. 107), with their accompanying explanation.

The remainder of the axial skeleton (fig. 108) consists of central and peripheral parts. The latter form consists of central and peripheral parts. The latter form the arches which surround the neural and visceral cavities. The former is constructed of separate pieces of bone (the *centra* of the vertebræ), connected by *intervertebral cartilages*. In the frog each centrum of a vertebra has its hinder end curved or rounded, to fit into a corresponding hollow in the anterior end of the succeeding centrum. Hence the articulation is said to be **procœlous**. The number of vertebræ never exceeds eleven—two forming the long peculiar coccyx, one the sacral bone, and eight the portion of the column anterior to the sacrum. The atlas or first vertebra has two hollow cups, which receive the *two* condyles of the skull; "but there is no specially modified axis vertebra." The transverse processes are very long, appearing as rudimentary ribs. Strong ligaments pass from spine to spine, and give strength to the vertebral column.

218. In mammals the under jaw-bone (mandibular *ramus*) is made up of one piece, and is directly articulated to the skull at the squamosal bone. Each side, however, of the mandible of the frog is composed of several pieces, and is connected with the skull by the intervention of the **quadrate bone** (*os quadratum*).

219. Appendicular Skeleton: the Limbs.—We now proceed to examine the appendicular parts of the skeleton. We observe, in the first place, that both hind and fore limbs may be divided into two parts:—1, a shoulder (or hip); 2, a limb proper. Or if we draw a horizontal line along the body, through the points at which the limbs proper are articulated to it, the part above this line will be the *dorsal* portion of the body; that beneath it will form the ventral portion.

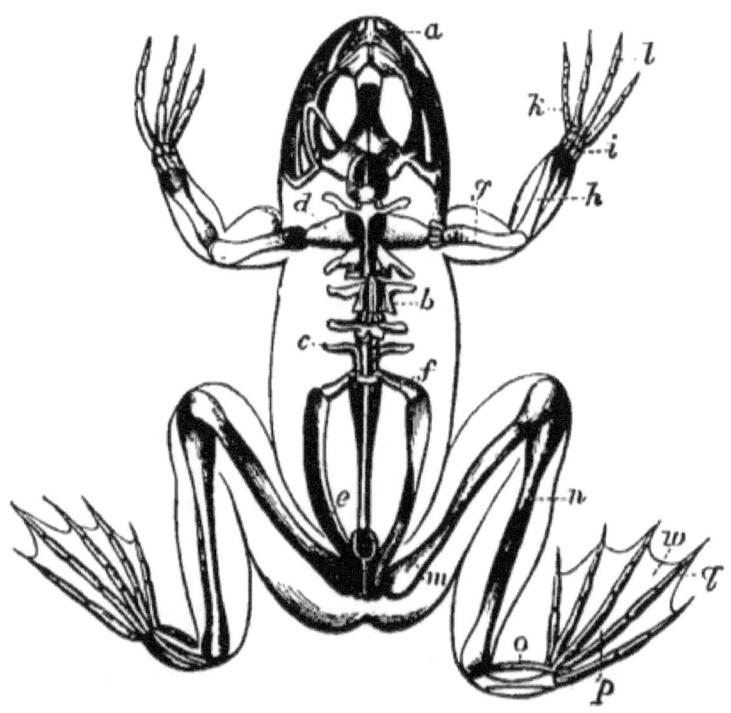

Fig. 108.—SKELETON OF FROG (after Nicholson). *a*, Skull; *b*, vertebral column; *c*, transverse process, or rudimentary rib; *d*, scapula, dorsal part of scapular arch; *e*, *os innominatum* of the pelvic arch; *f*, transverse process of single sacral vertebra; *g*, humerus; *h*, antebrachium, formed of ulna and radius fused into one bone; *i*, carpus; *k*, metacarpus; *l*, phalanges; *m*, femur; *n*, crus, formed of tibia and fibula ankylosed together; *o*, tarsus; *p*, metatarsus; *q*, phalanges; *w*, web.

220. The **pectoral arch** (fig. 108, *d*) is connected with the vertebral column by *ligaments*. The dorsal portion consists of the **scapula** *d*, which constitutes what is usually termed the "shoulder blade." The limb articulates with the pectoral arch at the **glenoidal cavity.** This is a hollow or socket, which receives the rounded end of the **humerus** or first bone of the fore limb. It is formed at the junction of the scapula and **coracoid** bone. The latter bone, with the **sternum** or "breast bone," forms the *ventral* portion of the pectoral arch. The sternum is the part in front into which the right and left coracoids become ankylosed. It also gives attachment to some of the ribs in all the air-breathing animals which possess them. The **clavicle** or "collar bone" (of which there is one on each side) also unites the scapula with the sternum, somewhat farther forward. It is a "membrane bone," and does not combine with the scapula and coracoid to form the glenoidal cavity.

221. The **humerus** *g*, or first bone of the fore limb, is jointed into the glenoidal cavity. Below, at the elbow joint, it is attached to the *ante-brachium* or fore-arm *h*, which consists of two bones placed side by side, but ankylosed into one. In man these bones, the *radius* and *ulna*, are separate, and movable on each other, and allow of a rotation of the hand, to affect what are called *pronation* and *supination*. Pronation is the normal condition of the *manus* in the frog; and no departure from this position is allowed, inasmuch as the ulna and radius in that animal are fused into a single bone. Attached at the lower end of the radius are the bones of the **carpus** *i*, then the four bones of the **metacarpus** *k*, corresponding to the four digits of the hand. Each **digit** or finger consists of three **phalanges** or finger joints *l*. The digits of the fore limb are not provided with a "web," as are those of the hinder limb.

222. The **pelvic arch** *e* is attached to the transverse processes of the sacral vertebra. In this it differs from the scapula, which is only attached by muscles to the vertebral column. Each side of the pelvic arch gets the name of **os innominatum** or "nameless bone." In the frog it is formed by the fusion together of two subordinate bones, the **ilium** and the **ischium**. A third bone, called the *os pubis* or pubic bone, which is found in higher vertebrates, is not represented in the amphibia. "The applied flat faces of the expanded ventral division of the pelvic arch coalesce into a disc."

223. The rounded head of the **femur** or thigh-bone *m*, is received into a cavity in the innominate bone, called the **acetabulum**, thus forming a "ball and socket" joint like that observed in the shoulder. At its lower end it is jointed to the **crus** or **tibia** *n*. There is a patella or knee pan covering this joint. The **fibula**, a smaller bone parallel to the tibia, is, in man, a separate bone; but in the frog it forms one bone with the tibia. The **tarsus** *o*, or small bones of the foot, come next; then the **metatarsus**, which gives origin to the digits. Four of the digits of the foot have each three phalanges; the remaining digit (the **hallux**) or "great toe," has only two. In the common frog, the spaces between the digits of the hinder limbs are "webbed."*

224. Physiology—Alimentation.—The frog has small teeth, arranged in concentric semicircles on the upper jaw, and on the vomerine plates in the palate. The tongue is attached, not to the hyoid bone behind, as in other vertebrates, but anteriorly to the "symphysis of the mandible," *i.e.*, the point of meeting of the under jaws. The *tip* is, therefore, turned backward; the *base*, forward. Living insects, the ordinary prey of the frog, are captured by rapidly darting forward the prehensile tongue. There are no salivary glands opening into the mouth of the frog. From the mouth the food passes into the **gullet** or **œsophagus**, whence it is transferred to the stomach, where it is acted upon by the **gastric juice**, a fluid secreted by numerous glandular follicles fixed in the walls of the stomach, and opening into its cavity. This

* Some species of frogs have no webs to their feet.

"juice" dissolves the nitrogenous parts of the food (the proteids), rendering them capable of being at once absorbed into the system, by means of the blood capillaries so largely distributed over the interior of the stomach and intestines. After leaving the stomach the food is exposed to the action of the **bile** and **pancreatic juice**, the former being secreted by the liver, a large gland which covers the stomach; the latter, by a second gland placed under the stomach, and termed the *pancreas*. The liver is constantly secreting bile from the blood. When no food is passing through the intestines, the bile is temporarily lodged in a receptacle called the **gall bladder**, which opens into the ordinary bile duct. It is only when food is passing along through the intestine that the gall bladder discharges its contents. The bile and pancreatic juice act upon the "fatty" parts of the food, turning them into what is called an *emulsion*. Thus prepared, the fatty materials are more readily **absorbed** into the lymphatic vessels which abound in the **villi** of the intestines, and which are called **lacteals**. The lymphatic vascular system is well developed in the frog. Its vessels convey *towards* the heart a fluid called lymph, which is derived partly from the food, partly from that portion of the blood which has exuded from the capillaries to bathe the tissues, and has not again been taken up by the veins. "Lymph may, in fact, be regarded as the blood *minus* its red corpuscles, and diluted with water, so as to be somewhat less dense than the serum of blood." The corpuscles in lymph are all of the colourless kind. The lymph is propelled forward by two pairs of contractile vessels or "lymphatic hearts;" one pair beneath the skin, immediately behind the hip joint, the other more deeply seated at the upper part of the chest. The intestine is simple and short, and terminates, not at the surface of the body, but in a **cloaca** or chamber, which also receives the renal excretions, and either the sperm cells or ova, according as the animal is male or female.

225. Respiration.—As the frog has no ribs, it cannot

respire by the alternate contraction and expansion of the cavity of the chest. The air passes in through the nostrils, which open into the cavity of the mouth, and are provided with valves opening *inwards*. When fresh air is required in the lungs, the frog, which habitually keeps its mouth shut, closes also the nasal valves, and by a movement of the hyoid bone, pumps the air down into the œsophagus. A slit in the under floor of the œsophagus, called the **glottis**, admits the air through a short laryngeal chamber (but without the intervention of a **trachea** or **bronchi**) into the lungs, among the air-cells of which (fig. 109) the venous blood is circulating. The animal could not succeed in thus *swallowing* the air, if either the open mouth or nasal passages allowed the air to stream out again. Hence a frog could be suffocated by keeping its mouth open.

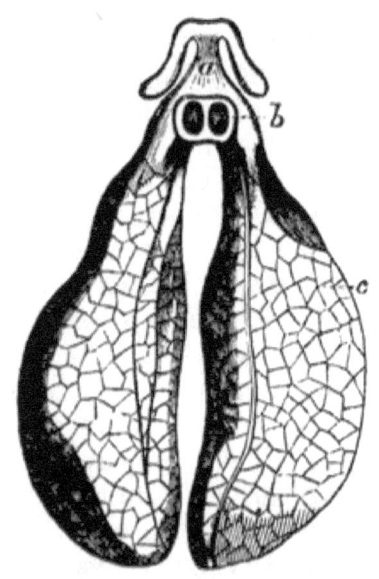

Fig. 109.—LUNGS OF A FROG (after Carpenter). *a*, Hyoidean apparatus; *b*, cartilaginous ring at the root of the lungs; *c*, pulmonary sacs, with reticulations marked out by cartilaginous framework.

The skin, too, of the frog being largely provided with capillaries, aids considerably in the aëration of the blood. In cold weather, when the flow of the blood is languid, this dermal respiration is sufficient to maintain the life of the animal. For respiratory purposes the skin must be kept moist. This is effected by numerous glands, which pour out their secretions on its surface.

226. Circulation.—In the anodon and lobster the heart is entirely *systemic*, *i.e.*, it is employed solely in propelling blood to the tissues of the body generally, after having received it from the respiratory organs, the gills. In the adult frog, the heart sends blood both to the general system to nourish the tissues and remove the waste products, and to the lungs and cutis to have it aërated, and converted from *venous* into *arterial* blood.

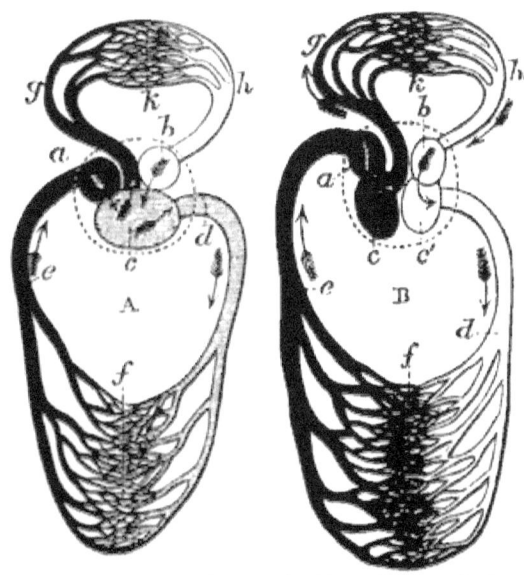

Fig. 110.—A, Plan of circulation in Reptiles and Amphibia. B, Plan of circulation in Birds and Mammals. *a*, Right auricle, receiving the systemic venous blood; *b*, left auricle, receiving bright "arterialised" blood from lungs; *c, c'*, ventricles. In A, where there is but one ventricle, the arterial becomes mixed with venous blood; in B, where the ventricle is double, they are kept distinct; *d, e, f*, systemic artery, vein, and capillaries; *g, h, k*, pulmonary artery, vein, and capillaries. The white, black, and shaded spaces denote "arterial," "venous," and mixed blood, respectively. The arrows indicate the direction of the current.

There are, as in the anodon, and in most reptiles, *two* auricles, and *one* ventricle (fig. 110, A, *a, b, c*). The vessels *d, g*, which carry blood away *from* the heart, pass out from the ventricle, and are called **arteries**. These branch off into smaller and smaller arteries, and ultimately into very fine tubes *f, k*, called **capillaries**, with thin structureless walls. The **pulmonary** capillaries *k* ramify through the membranous tissues of the lungs; the **systemic** capillaries *f* supply blood to the tissues of the body generally. The blood, after flowing through the capillaries, is collected into **veins** *e*, *h*, which carry it *towards* the heart; the systemic blood to the *right*, the pulmonary to the *left*, auricle. The flow of blood through the respiratory organs is called the *pulmonary* or lesser circulation; that through the body generally is termed the *systemic* or *greater* circulation. In reptiles and amphibia, the blood of both circulations mingles in the ventricle. In birds and mammals, which have *two* ventricles (fig. 110, B), one for the pulmonary, the other for the systemic, circulation, there is no direct communication between the two sides of the heart, and the two circulations are kept quite distinct from each other. The blood in its course does not, in vertebrates, collect in cavities or lacunæ, hollowed out in the tissues, as in invertebrate animals, but courses through vessels (arteries, capillaries, and veins) with definite wall structures.

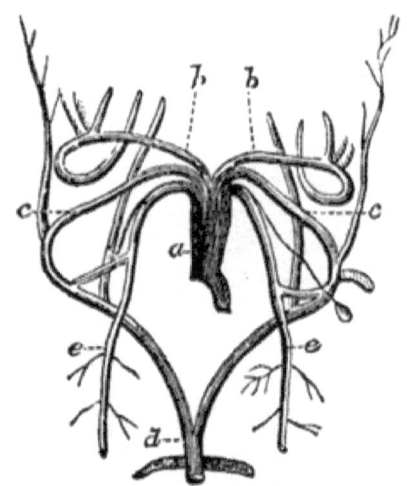

Fig. 111.—THE ARTERIAL SYSTEM IN AN ADULT FROG. *a*, Bulbus arteriosus; *b*, *b*, carotid arteries, carrying bright, arterial blood to the head; *c*, *c*, aortic arches, uniting at *d*, in the dorsal aorta, the great systemic artery; *e*, *e*, pulmocutaneous arteries, carrying venous blood to the lungs and skin to be aërated.

227. In the frog, the blood contained in the ventricle is forced, by the contraction of that cavity, into a chamber (also rythmically contractile) called the **bulbus arteriosus** (fig. 111, *a*), whence it passes into either of three pairs of large curving arteries, or *aortic trunks*, *b*, *c*, *e*. The first pair of aortic trunks form the **carotid arteries** *b*, *b*,

and carry the blood to the head. The second, *c, c*, unite, after leaving the heart, into one trunk, the **dorsal aorta**. The third or hindmost pair, the pulmo-cutaneous arteries, *e, e*, convey venous blood to the lungs and to the integument, to be there aërated while passing through their capillaries. The pulmonary *veins* carry the oxygenated blood to the *left* auricle (fig. 110, A, *b*), whence it passes into the ventricle. The left auricle in amphibia is much smaller than the right, and admits the oxygenated blood (which is of a much brighter hue than venous blood) by a single pulmonary vein. The blood of the systemic capillaries supplies fresh materials to the tissues to which it is distributed, and oxygen also, to assist in the decomposition of the decaying portions of those tissues. The oxidated waste matters are generally resolved into *carbonic acid*, *water*, and *urea* or *uric acid*, and are carried off in the current of the blood, whence they are removed by the excretory organs of the body, namely, the *lungs*, the *skin*, and the **Wolffian bodies**, these last being the representatives of the *kidneys* of the higher vertebrates. The walls of the capillaries are very thin structureless membranes, which allow of the transfusion through them of the fluid portions of the blood. After traversing the tissues, the capillaries unite again into larger trunks, called **veins**. The smaller veinlets again combine into larger veins, which convey the blood ultimately into a pulsating chamber connected with the right auricle, and called the *sinus venosus* (fig. 112, *b*). From this chamber the venous blood, which is of a dark red hue, passes into the right auricle, whence it is received into the ventricle. The latter chamber also receives "arterialised" (*i.e.*, *oxygenated*) blood from the left auricle (fig. 110, *b*). Hence

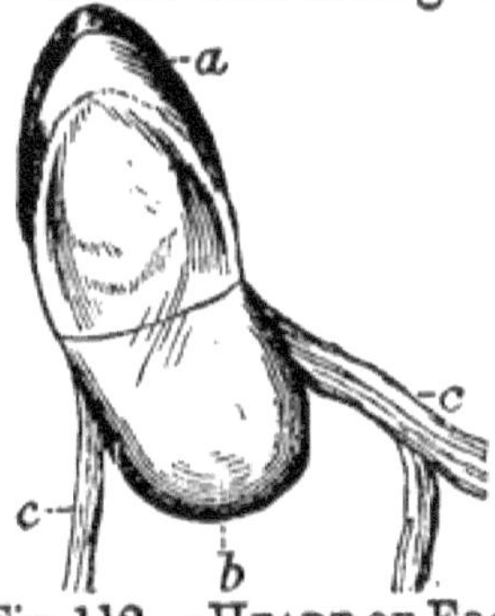

Fig. 112.—HEART OF FROG turned upward to show sinus venosus behind. *a*, Heart; *b*, sinus venosus; *c*, veins.

the blood in the ventricle is a mixture of *venous* and *arterial* blood. " Only venous blood passes into the pulmonary arteries of a frog, while mixed blood enters the aortic arches, and is of a brighter arterial hue at the end, than at the beginning, of the systole. The blood in the carotid passages is always bright."* Hence the head, including the brain, is always supplied with pure arterial blood. The mechanism by which this singular result is brought about is too complicated to be described in an elementary treatise like the present. As a consequence of the imperfect oxygenation of the blood in amphibia, reptiles, and fishes, its temperature rises but little above that of the surrounding medium. Hence these are termed *cold-blooded animals.* The red corpuscles in the blood of the frog are large, oval, and nucleated, and their passage through the capillaries of the web of the foot, as observed under the microscope (fig. 113), furnishes a very beautiful and interesting spectacle.

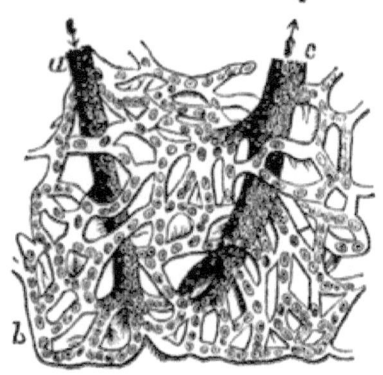

Fig. 113.—CIRCULATION IN A FROG'S "WEB," as seen under the microscope. *a*, Small artery; *b*, capillaries; *c*, veinlet. The red corpuscles are distinctly visible coursing through the capillaries. The arrows show the direction of the current.

228. **Epidermis and Epithelium.**—It has already been observed that connective or fibrous tissue, cartilage, and bone, are formed by the differentiation of the protoplasmic matter of certain nucleated cells. We have now to add that the remaining tissues of the body are formed either entirely of such cells, or else of structures manufactured from them, by the active principle contained in the protoplasm. Epidermis, epithelium, scales, claws, and other horny tissues consist, like the parenchyma of

* Huxley.

plants, entirely of cells. If a vertical section of **epidermis** (or cuticle) be examined under the microscope, the deep-seated parts are seen to be formed of spheroidal cells, filled with fluid, and containing a nucleus. These absorb nutriment from the blood which circulates in the connective tissue of the **dermis** beneath; they add to the thickness of their cell walls, and, as they grow older, are pushed outwards by the development of new cells beneath. These, in their turn, are also pushed outwards by a still younger layer of cells. As the cells approach the surface they become gradually flattened, and when they arrive there it is observed that their protoplasm has disappeared, leaving nothing but flattened scales, which are ultimately cast off as dead matter. **Epithelium** is the cellular tissue which lines the internal cavities and the various *glands* of the body. In structure it differs from epidermis in never becoming dry or horny; but it grows, and is shed, in the same way. The epithelium cells of glandular structures, when they reach the free internal surface, become *ruptured*, and their contents flow into the *duct*, forming part of the peculiar secretion of the gland. In this way the gastric and pancreatic juices, saliva, bile, arachnoid fluid, etc., are generated.

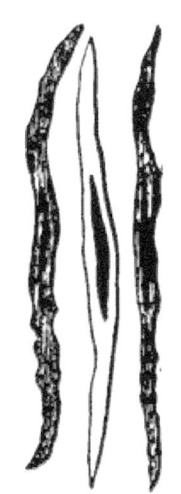

Fig. 114.—ELONGATED CELLS, constituing the fibre of smooth muscle. The middle one, treated with acetic acid, shows distinctly the rod-like nucleus.

229. **Muscular Tissue.**—The various movements of the body, or its parts, are effected by the contraction of numerous fibres called *muscles*. Muscular fibre is endowed with the property of contractility when acted upon by a stimulus, such as nerve force, contact, heat, cold, electricity, etc. The muscles of the frog are of two kinds, namely, **smooth** and **striated** (or *striped*). Smooth

or non-striated muscle is formed of bands of spindle-shaped cells (fig. 114) securely attached to each other. Each cell has a rod-like nucleus. On the application of a stimulus the cell contracts, *i.e.*, becomes shortened; but as it is filled with fluid, and is hence always of the same size, it is plain that when a muscle becomes shortened it must increase in thickness. Smooth muscular fibre constitutes the fabric of what are called the "hollow muscles" of the body, the heart alone being excepted. In the walls of the intestinal canal, the stomach, the principal gland ducts, and all but the smallest arteries, they are arranged in two thin layers; one longitudinal, by which the part of the organ where they act is shortened; the other transversal, by which it has its calibre narrowed. In the iris of the eye, the two sets are respectively *circular* and *radial*. The former set contract the pupil, the latter cause it to dilate.

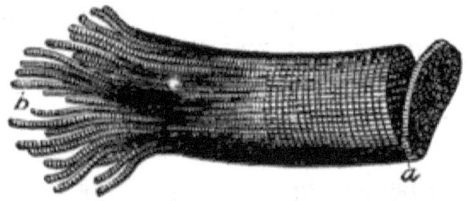

Fig. 115.—STRIPED MUSCULAR FIBRE, breaking up at *a*, into transverse discs; at *b*, into fibrillæ.

230. **Striated** or **striped** muscle (fig. 115) is so named from its presenting cross markings of alternately light and dark shading. Fibres of striated muscle are not formed into flat sheets or bands, like smooth muscle, but into cylindrical or prismatic bundles, enclosed in a dense sheath of connective tissue, called a **fascia**. Each fibre, when further examined, is observed to be composed of a number of minute **fibrillæ**, also striated, and arranged parallel to each other, each collection being held together by a very thin sheath of structureless membrane, called a **sarcolemma**. A fibre, when torn, may resolve itself into fibrillæ, or may separate into transverse discs, as seen in the figure. Among the fibrillæ are found, scattered over the interior surface of the sarcolemma, a number of *nuclei*. From these appearances Professor Kölliker believes that

each fibre is formed from a single cell, by elongation, division of the nucleus, and differentiation of the protoplasm into fibrillæ. Striated muscle contains about 75 per cent. of water. After death the fluids in the fibres *coagulate*, rendering the muscles rigid. This is called *rigor mortis*, or the "death stiffening."

231. According to their *function*, muscles may be separated into two classes:—

1. Those not connected with bones or cartilages (hollow muscles).
2. Those which are employed to act upon *joints*.

In the former class are to be reckoned the heart, the muscles of the intestines, of the arteries, of the iris, of the gland ducts, and, in general, those muscles which are termed *involuntary*, or over which, when excited by a stimulus, the *will* has little or no control. To the second class belong all those which are ordinarily directed or controlled by the will. It is important to bear in mind that —the heart alone excepted—all the involuntary muscles in *vertebrate* animals are of the *smooth* or non-striated kind, while all the voluntary muscles are striated. It has been already observed that in the mollusca *all* the muscles are of the type called smooth or unstriped, while in the arthropoda they are *all* of the striated kind; so that the distinction which obtains in vertebrate animals is by no means universal.

232. A nervous impulse, either from the brain, spinal cord, or sympathetic nerve system, is the ordinary stimulus which directs the movements of the muscles. "The muscles are, so to speak, the servants of the nerves, *doing*, with a force of their own, the work which the nerves direct. The relation between the two may be likened to that of the rider and his horse, or of the engine-driver and his locomotive; for the nerves can put forth no motive power by themselves, whilst, on the other hand, the muscles (with certain exceptions) remain inert except when stimulated to contract by the agency of the nerves."*

* Carpenter.

233. The Nervous System.—The **cerebro-spinal axis** or nerve-system occupies the neural chamber of the vertebral column, including the cavity of the skull. All this line of nerve matter is but loosely connected with the sides of the bony cavity, in which it lies, by means of the nerve-trunks that pass outward through what are called the inter-vertebral foramina. A stout fibrous tissue lines the inside of the neural canal throughout its whole length. This lining corresponds in its functions with the *periosteum* which invests ordinary bone. It is formed of connective tissue, and is named the **dura mater**.

234. The substance of the nervous tissue is made up of nerve cells, nerve fibres, and a *framework* of connective tissue. This tissue is better developed on the external surface of the cerebro-spinal axis than elsewhere, and here also it is largely supplied with blood-vessels; but as it enters into the substance of the nervous matter, it can scarcely be termed a *coating* or *investment* of the cerebro-spinal axis. As it is more delicate and slight than the membrane which lines the cavity, it is called the **pia mater**. Both the dura mater and the pia mater pass into a delicate connective tissue, lined with epithelium, to which the name of **arachnoid membrane** is given. It forms a closed sac or cavity, whose epithelial cells secrete a fluid which fills the arachnoid cavity, and which *eases* the friction of the cerebro-spinal axis upon the wall of that cavity, in the same manner as the serous fluid of the pericardial sac eases the movements of the heart. If the cerebro-spinal axis *actually touched* the wall of its cavity, serious

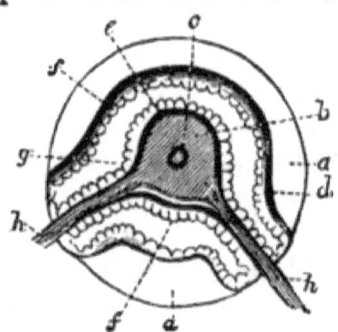

Fig. 116. — DIAGRAMMATIC VIEW OF THE LINING MEMBRANES OF THE CEREBRO-SPINAL AXIS. *a, a,* Casing of bone ; *b,* nerve matter ; *c,* central canal ; *d,* dura mater; *e,* pia mater; *f, f,* arachnoid membrane (epithelium); *g,* arachnoid cavity; *h, h,* roots of nerves.

injury to the health of the animal would ensue. The arachnoid cavity is closed, and though nerves are always passing outwards at intervals, the membrane invests them so closely that there is no real opening for the cavity itself (fig. 116).

235. The Spinal Cord (fig. 117, *k*) extends as far as the sacrum, after which it exists only as a fine filament. At the sacrum a number of *nerves* pass out, and form a *plexus* to supply the hinder limbs. In a general way, it may be observed, that a pair of nerves (one at each side) pass out between every two vertebræ of the spinal column. These are called *spinal* nerves; others pass out through the skull, and are hence called *cerebral*.

236. If a transverse section be taken of the spinal cord (fig. 118, A), it will be found that it is deeply penetrated by two narrow clefts,

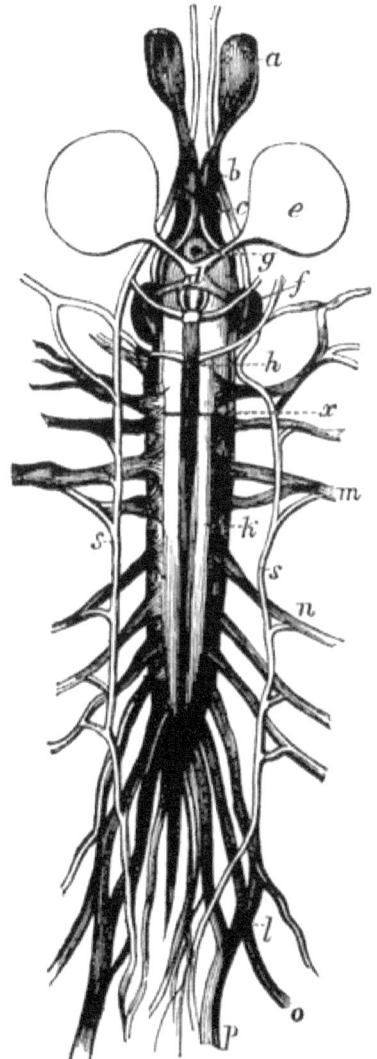

Fig. 117.

Fig. 117. — DIAGRAM OF THE CEREBRO-SPINAL AXIS, SPINAL NERVES AND SYMPATHETIC SYSTEM OF A FROG, viewed from below. At *x* is shown the boundary line between the brain and spinal cord; *a*, olfactory sac; *b*, olfactory lobe; *c*, cerebral hemisphere; *d*, optic chiasma; *e*, eye; *f*, optic lobe; *g*, pituitary body; *h*, medulla oblongata; *k*, spinal cord; *l*, ischiatic plexus; *m*, brachial plexus; *n*, spinal nerve; *o*, crural nerve; *p*, ischiatic (or sciatic) nerve; *s, s*, sympathetic nerve chains.

behind and before, *a*, *b*, almost meeting in the middle, the section showing two semicircles, united by a "bridge" of nerve matter. The two clefts are termed the *anterior* and *posterior fissures*. The latter is much deeper than the former, and a fold of the *pia mater*, with its blood vascular system, dips to the bottom of each, thus supplying blood to the vesicular or "grey matter" *d* of the interior. The spinal cord may hence be regarded as formed of two half cylinders, with their flat faces apposed, but adhering to each other near the middle of their line of junction. This "bridge" of nerve matter, which unites the right and left sides of the cord, encloses a small circular canal, which is lined with epithelium, and extends throughout the whole length of the cerebro-spinal axis. It is called the **central canal**. The cross section of the cord shows also a distinction of its substance into "white matter" and "grey matter." The latter is internal, and appears as a pair of crescent-shaped patches, arranged "back to back," one on each side, with the convex borders united across the "bridge" by an intervening mass of nerve matter which is partly

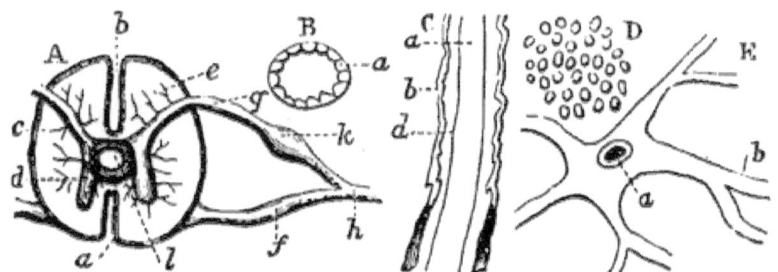

Fig. 118.—A, Transverse Section of Spinal Cord: *a*, anterior fissure; *b*, posterior fissure; *c*, lateral hemicylinder; *d*, "grey" matter; *e*, white matter; *f*, anterior, *g*, posterior, roots of nerve trunk, *h*; *k*, ganglion of posterior root; *l*, central canal. B, Central canal, enlarged to show *a*, epithelium lining. C, Nerve fibre: *a*, axis cylinder; *b*, sheath; *d*, white substance of Schwann. D, Cross section of white matter, showing the cut ends of nerve fibres. E, A ganglionic corpuscle: *a*, its nucleus; *b*, one of its prolongations.

vesicular and partly fibrous. Their concave outlines are directed outwards, and their anterior and posterior *cornua* are connected with the respective roots of the spinal nerves. The "white matter" *e* is external to the "grey" or vesicular matter; and when examined under the microscope, it is found to consist of a collection of minute tubular fibres, laid side by side, of which a cross section is given in fig. 118, D. The grey or vesicular matter is formed of a mass of ganglionic corpuscles or nerve cells (fig. 118, E). The corpuscles are nucleated, and of a *stellate* form. They give off prolongations, some of which unite with other corpuscles; and some pass into nerve fibres.

237. Each nerve has two "roots" (fig. 118, A, *f*, *g*), connected respectively with the anterior and posterior "cornua" of the crescent of grey matter; but most of the fibres of which they are constituted have their origin in the brain, and form a component part of the spinal cord, until they pass out from it as *nerves*. Each "root" is made up of a bundle of nerve fibres, enclosed in an envelope or sheath of connective tissue, called a **neurilemma**. The posterior root is provided with a **ganglion** or enlargement *k*, containing nerve corpuscles. The anterior root has no ganglion. Both roots unite to form a *nerve trunk h*, which passes out from the neural canal through an aperture between two vertebræ, hence called an **inter-vertebral foramen**. From the same point in the axis two nerve trunks issue, one proceeding to each side of the body.

238. After passing into the common nerve trunk, there is still a distinction in function between the nerve fibres of the two roots, though nothing in their *appearance* indicates to which root any of them may belong. If the spinal cord of an animal be laid bare, and the posterior root of a nerve be irritated in any way, violent pain will be felt, but no involuntary movement will take place. If, on the other hand, the anterior root be irritated, violent convulsive movements will be observed in the part of the body in which the nerve terminates, but

no pain will be felt. By cutting the posterior root, and irritating each part, it will be found that its nerve fibres are **afferent**, that is, they carry nervous impulses *towards* the spinal cord and the brain, but not in the other direction. But if, instead, the anterior root be divided, it will be found that its fibres can only convey impulses *outwards* or *from* the spinal cord, towards the muscles whose movements they direct. Hence these are called **efferent** nerves. It is thus seen that the nervous impulse which gives rise to a sensation of any sort, such as sight, sound, taste, smell, touch, or the like (which is always felt in the brain, though referred to that part of the body to which the nerve is distributed), must pass *inward* and *to* the brain by the *posterior* or gangliated root; while the nervous impulse, by which the will directs a muscular movement, passes *outward* and *from* the brain by the *anterior* root. An afferent nerve, therefore, in full communication with the brain, will be **sensory**, and an efferent acting upon a muscle will be a **motor** nerve.

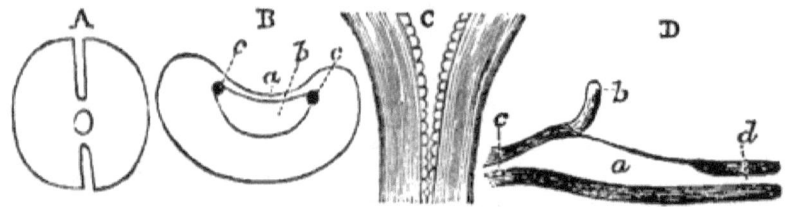

Fig. 119.—A, Section of medulla oblongata, near posterior end. B, Do., at anterior end: *a*, cavity corresponding to posterior (or dorsal) fissure; *b*, cavity of fourth ventricle, being a widening of central canal; *c*, "restiform bodies" which give support to the cerebellum. C, Longitudinal and horizontal section of medulla oblongata, showing the widening of the central canal, to form the cavity of the fourth ventricle. D, A vertical longitudinal section of same, to show cerebellum and fourth ventricle: *a*, fourth ventricle, *b*, cerebellum: this organ is very small in the frog; *c*, *iter a tertio ad quartum ventriculum;* *d*, central canal.

239. The Brain (or **Encephalon**).—The **medulla oblongata** (fig. 117, *h*) is the anterior portion of the spinal cord, and is contained within the cavity of the skull. A transverse section (fig. 119, A) near its *posterior* end differs but little from that of the spinal cord already given (fig. 118, A). A cross section near its *anterior* end is given in fig. 119, B. In comparing it with that shown at A, it will be seen that the anterior fissure has disappeared, and that the remaining one is turned into a simple concavity *a*. The central canal, too, has changed its shape, become wider, and approached the concave side, from which it is separated by a very thin membrane, "which tears with great readiness, and lays open the cavity of the fourth ventricle," as the expanded canal is now named. This membrane is not formed of *nerve tissue*, but of the *ependyma* or lining of the brain cavity, and of the arachnoid membrane, which covers the brain externally. Fig. 119, D, gives a vertical and longitudinal section of the medulla oblongata, and the **cerebellum** or *little brain*. This is a part that has been superadded to the previously existing neural axis. Fig. 119, C, gives a horizontal and longitudinal section of the medulla oblongata as it opens into the fourth ventricle. The central canal becomes widened, and is observed to be still lined with epithelium.

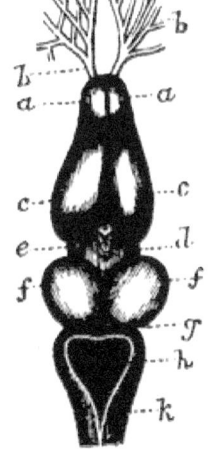

Fig. 120.—BRAIN OF FROG, as seen from above. *a, a*, Olfactory lobe; *b, b*, olfactory nerves; *c, c*, cerebral hemispheres; *d*, thalamencephalon, enclosing third ventricle; *e*, pineal gland; *f, f*, optic lobes, or "corpora bigemina"; *g*, cerebellum; *h*, fourth ventricle; *k*, medulla oblongata.

240. The axis continues onward, but the central canal becomes narrowed into the "*iter a tertio ad quartum ventriculum*" (or passage from the third to the fourth ventricle, fig. 121, *f*). Over this passage are two rounded masses, called the **optic lobes**, or *corpora bigemina* (fig. 121, *l*, and fig. 120, *f, f*). At the anterior end of the

passage is the third ventricle (fig. 121, *d*), resulting from a vertical expansion and lateral compression of the central canal. From the upper external surface of this part of the brain, a little cylindrical or sub-conical mass (fig. 121, *i*, and fig. 120, *e*), called the *pineal gland*, projects, while the floor of the ventricle is produced into a hollow conical process, called the **infundibulum**, to the under end of which is attached a rounded constricted mass called the **pituitary body** (fig. 117, *g*, and fig. 121, *p*).

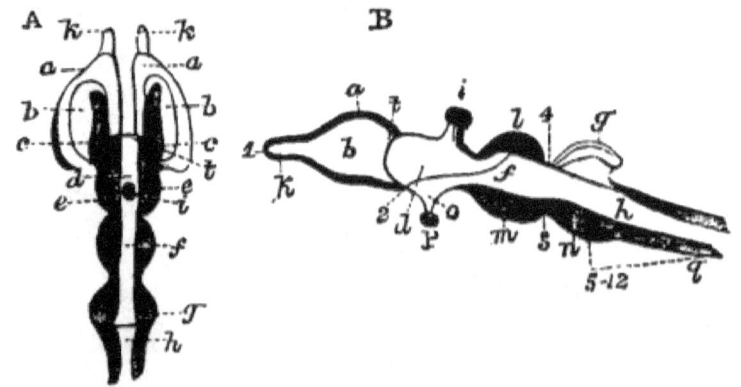

Fig. 121.—Diagrammatic Longitudinal Sections (A, horizontal; B, vertical) of the Brain of a Typical Vertebrate Animal. *a, a*, Cerebral hemispheres; *b, b*, lateral ventricles; *c, c*, corpora striata; *d*, third ventricle; *e, e*, optic thalami; *f, iter a tertio ad quartum ventriculum; g*, cerebellum; *h*, fourth ventricle; *t*, terminal lamina; *i*, pineal gland; *k, k*, olfactory lobes; *l*, optic lobe (*corpora bigemina*, aut quadrigemina); *m*, crura cerebri; *n*, pons Varolii; *o*, infundibulum; *p*, pituitary body; *q*, medulla oblongata. Some of these parts are only rudimentary, or even absent, in the frog. The figures denote the points whence the nerves, so numbered, originate.

The roof of the third ventricle is, like that of the fourth, a mere membrane. The anterior boundary of the third ventricle is called the **lamina terminalis** (fig. 121, *t*). But though the central canal terminates here in the middle line, it projects forward in two horn-like branches, one into either **cerebral hemisphere**. These cavities are called lateral ventricles (fig. 121. *b*). The cerebral hemispheres in man are very large, and extend backward so as to almost cover the cerebellum. They also project over the olfactory lobes in front. In the frog, however, the hemispheres (fig. 120, *c, c*) are not nearly so well developed. Anterior to the hemispheres are two little projecting bodies called the olfactory lobes (fig. 120, *a, a*, and fig. 121, *k, k*), because they give off minute filaments (fig. 120, *b, b*) to the organs of smell. That portion of the brain which includes the third ventricle and the parts in front of it, is called the **fore-brain**; the part enclosing the passage from the third to the fourth ventricle constitutes the **mid-brain**; while the **hind-brain** is made up of the medulla oblongata and the cerebellum, and includes the cavity of the fourth ventricle. In the spinal cord and medulla oblongata the white matter is external, and the grey internal. The reverse arrangement holds in the cerebellum and cerebral hemispheres.

241. Cerebral nerves are those that issue directly from some part of the brain. They are arranged as follows, in *ten* pairs, one trunk going to each side :—

Name.	Character.	Origin.	Distributed to
I. Olfactory.	Sensory.	Olfactory lobes.	Nasal chambers.
II. Optic.	Sensory.	Optic lobes.	Organs of sight.
III. *Motores oculorum.*	Motor.	Crura cerebri.	Muscles of eye.
IV. *Pathetici.*	Motor.	Between optic lobes and cerebellum.	Superior oblique muscles of eye.
V. Trigeminal.	Sensori-motor.	Medulla oblongata.	Integument of head (sensory part). Muscles of jaws (motor part).
VI. *Abducentes.*	Motor.	Medulla oblongata.	External rectus muscles of eye.
VII. Facial.	Motor.	Medulla oblongata.	Facial muscles.
VIII. Auditory.	Sensory.	Medulla oblongata.	Organs of hearing.
IX. Glossopharyngeal.	Sensori-motor.	Medulla oblongata.	Tongue (sensory part). Muscles of pharynx (motor part).
X. Pneumogastric or *par vagum.*	Sensori-motor.	Medulla oblongata.	Gullet, stomach, respiratory and vocal organs, heart, and integuments.

The *spinal accessory* and *hypoglossal* nerves—being respectively the eleventh and twelfth pairs in the higher vertebrates—are not met with in fishes and amphibia.

242. The Sympathetic Nervous System.—Besides the cerebro-spinal system of nerves, most vertebrates (all, indeed, but a few fishes) have in addition what is termed the *sympathetic system*. This consists (fig. 117, *s, s*) of two chains of ganglia in front and on either side of the spinal column. These ganglia are, however, connected with the cerebro-spinal system, receiving into their mass some of the spinal nerves of their respective sides. Fibres from the sympathetic system accompany the blood-vessels, and regulate the quantity of blood which is to pass through them, by narrowing or widening their calibre. Sympathetic nerves are also employed to govern and direct the muscles of the intestines. The heart, too, and the glands or secretive organs of the body, perform their ordinary functions under the control of nerves passing into them from the sympathetic system. These nerves may be either afferent or efferent; but the afferent nerves are not *sensory*, and the efferent or motor nerves are not under the control of the will.

243. The Sensory Organs.—" The organs of the three higher senses—smell, sight, and hearing—are situated in pairs upon each side of the skull in all vertebrate animals except the lowest fishes; and, in their earliest condition, they are alike involutions of the integument."*

(*a.*) The nasal sacs form the organs of **smell**. Nerve filaments from the olfactory lobes are distributed over the mucous lining of these chambers, and convey to the brain the impression created by the *contact* of very minute particles of the substances from which the odours arise.

(*b.*) The **eye** or organ of *sight* " is formed by the coalescence of two sets of structures, one furnished by involution of the integument, the other by an outgrowth of the brain."* The structure of the eye will be better understood from a diagram than from a written description.

* Huxley.

THE SENSORY ORGANS OF FROG.

The sensation of vision is produced by rays of light passing through the cornea (fig. 122, *b*), aqueous humour *m*, crystalline lens *g*, vitreous humour *l*, and finally impinging on the retina or nervous matter of the eye at *e*. In consequence of their successive refractions in their passage through these media, the rays come to a focus on the retina, forming there an *inverted* image of the object, just as they would in an ordinary *camera obscura*. The frog has two eyelids, and a *nictitating* membrane. The "Harderian gland" pours out its secretion on the eyeball.

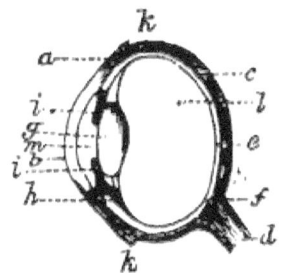

Fig 122.—HORIZONTAL SECTION OF EYEBALL. *a*, Sclerotic coat; *b*, cornea; *c*, the choroid coat, containing a black pigment; *d*, the optic nerve; *e*, the retina; *f*, the blind spot; *g*, the crystalline lens; *h*, ciliary process; *i, i*, iris; the part of the crystalline lens uncovered by the iris is the *pupil;* *k, k*, points of attachment of the recti muscles; *l*, vitreous humour; *m*, aqueous humour.

(*c.*) The sensation of *sound* is conveyed to the brain by means of the auditory organs, the **ears**. Each ear consists (1) of a cavity called the **tympanum**, communicating freely with the throat, and covered externally by a membrane called the tympanic membrane; and (2) of a **labyrinth**, consisting of three *semicircular canals*, surrounded by the bony structures of the skull. To the inner side of the tympanic membrane is attached the outer end of a little bone, called the **stapes** or *columella auris*, the other extremity of which is similarly fixed to an inner membrane called the **fenestra ovalis**. The latter membrane separates the tympanum from the labyrinth. Each of the semicircular canals is filled with a fluid called *endolymph*, containing little granules of carbonate of lime, called **otoliths**. The internal surface of the canals is overlaid with filaments from the auditory or eighth pair of cerebral nerves. Surrounding the membranous canals is the *perilymph*. Sound carried to the ear causes the tympanic membrane to vibrate; the vibration is communicated to the stapes, which imparts it to the fenestra ovalis. This, in its turn, communicates the vibratory movements to the endolymph, and its contained otoliths. These striking rythmically on the auditory nerve, distributed over the interior of the labyrinth, impart to it the impression of sound, which the nerve at once conveys to the brain.

The *cochlea, malleus, incus*, and *external ear* of mammals are wanting in the frog.

244. Reproduction.—The eggs are deposited in the water, and may there be observed strung together in masses, being attached to each other by their glutinous envelopes. In the centre of each is seen a dark speck— the **vitellus or yelk**. The ova are fecundated after their extrusion from the body of the female. The ciliated sperm corpuscles of the male enter the ova, become embedded in their vitelline masses, and the two protoplasms become thoroughly commingled. A series of remarkable changes are then set up in each ovum; the yelk divides into two masses, these again subdivide, and so the operation proceeds till the original vitellus assumes the appearance called the "mulberry mass," consisting of an aggregation of round nucleated cells. This process is called the "cleavage" or "segmentation" of the yelk; also, "endogenous cell multiplication." The cells, as they are produced, approximate towards the surface, where they form two layers, the outer called the **epiblast**, the inner the **hypoblast**, and *both* together constituting the **blastoderm**. These two layers correspond to the ectoderm and endoderm of the hydra, and like them, too, the outer coat consists of smaller cells closely packed together, while the inner is made up of larger cells, more loosely arranged. An oval spot appears on the surface, called the "germinal area" or "embryonic spot" (fig. 123, A). The marginal part of this spot is opaque, but it encloses a clear space called the "area pellucida" *a*, marked longitudinally by a hollow streak called the **primitive groove**

b. This groove becomes deeper by the rising upwards of its edges, formed of folds of the blastoderm (fig. 123, B). These folds, called **laminæ dorsales**, approach each other over the groove (C), and at last meet and coalesce (D), forming a tube which ultimately becomes the neural canal (E, *a*). The cells of the epiblast contained within the tube become, by differentiation, the substance of the cerebro-spinal axis of the nervous system. Simultaneously with this development, a long segmented cord (the **notochord** or **chorda dorsalis**, E, *b*) appears directly under the neural canal, which becomes the foundation of the future vertebral column. The notochord becomes enveloped in a fibrous sheath; rings, called *protovertebræ*, surround it, to form the bases of the centra, and send projections upwards into the dorsal laminæ to form the *neural arches*.

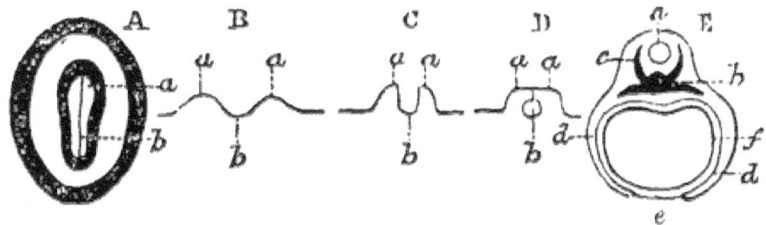

Fig. 123.—A, *a*, *area pellucida;* *b*, primitive groove. B, C, D, Cross sections of groove, showing the successive stages in the development of the *laminæ dorsales*, and the formation of the neural canal: *a, a*, dorsal laminæ; *b*, neural cavity. E, More advanced stage: *a*, neural cavity ; *b*, notochord ; *c*, neural arches; *d, d*, ventral laminæ approaching each other at *e; f*, hypoblast.

245. While these changes are in progress, the outer margins of the embryonic spot extend themselves laterally and downwards along the surface of the ovum (E, *d, d*), till they at length meet in the middle line (E, *e*), directly opposite to the neural canal, thus enclosing the whole vitellus. The latter is then gradually absorbed into the substance of the parts already formed. In the course of the development of the epiblast and hypoblast, these two structures form a layer between them, called the **meso-**

blast. Subsequently this layer splits into two separate sheets "throughout the regions of the thorax and abdomen, from the ventral margin nearly up to the notochord." The inner of these sheets remains in close connection with the hypoblast, forming with it the structure that ultimately becomes the wall of the intestinal canal. The other, united with the epiblast, becomes developed into the general body wall of the chest and abdomen.* In mammals, birds, and reptiles, there are many more stages in the development of the embryo, previous to its leaving the egg.

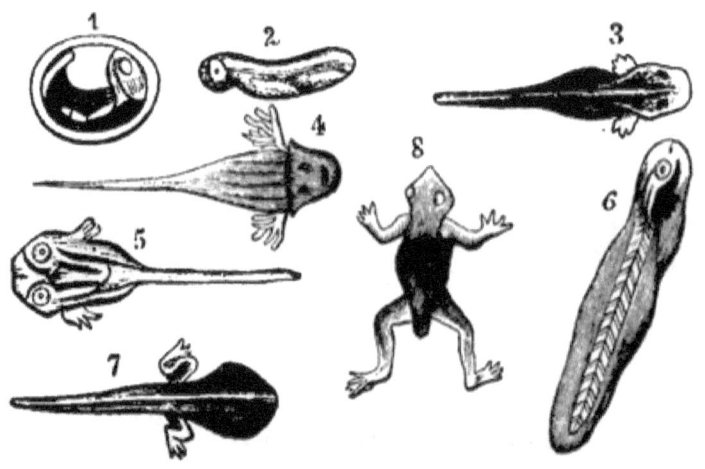

Fig. 124.—SUCCESSIVE STAGES IN THE DEVELOPMENT OF A LARVAL FROG.

246. The **Tadpole** or *larval frog* (fig. 124) comes forth from the egg in an imperfect condition, but provided with a large laterally compressed tail, by which it is able to dart rapidly through the water. The intestine is very long, and coiled up within the body like a watch spring.

* From the *epiblast* are formed the epidermis and glands of the skin, the cerebro-spinal *axis*, and the organs of sight and smell. The mesoblast forms the skeleton, muscles, connective tissue, vascular system, and the cerebral and spinal *nerves*. From the hypoblast are produced the epithelium of the thorax and abdomen, with their glandular system.

A horny beak is formed upon the mouth, and a pair of "suckers" appear under it, by which the little animal is enabled to take up a fixed position. When it leaves the egg it has no respiratory organs, but external ciliated gills, formed into large tufts at the rear part of the head, soon make their appearance. Each pair of gill tufts is at first naked, but an outgrowth of the integument expands over the branchial cleft, and eventually covers the gills, leaving only a small aperture through which they protrude. This aperture remains longer on the left than on the right side. A second set of gills then become developed beneath the operculum or cover, already referred to. These are formed of short filaments, and are attached to four pairs of branchial arches. The aërating water passes in by the mouth, bathes the gills, and passes out by one or two slits behind the operculum. In this condition the structure and appearance of the tadpole resemble those of a fish. The heart, which at this stage is entirely branchial, containing only venous blood, has but one auricle and one ventricle. The anterior limbs are first developed *under the operculum;* but on this account they are not visible till after the appearance of the posterior limbs. While these changes are in progress, the horny beak falls off; the blood is gradually diverted from the gills to the lungs; the gills in consequence become atrophied, and disappear; a *septum* is developed, dividing the auricle into two chambers; the intestine becomes much shorter as the animal becomes carnivorous; finally, the tail entirely disappears, and the tadpole takes the *form* of an adult frog.

247. Distinctive Characters of Vertebrate Animals.

1. A cross section of a vertebrate animal shows that the body consists of *two* tubes, the *neural* and *visceral*, completely separated from each other; while a similar section of an invertebrate discovers only *one* tube, the visceral.

2. In the vertebrate embryo, a *chorda dorsalis*—generally replaced in the adult by the vertebral column—

separates the neural from the visceral cavity. No such structure is found in invertebrata.

3. The "visceral arches" of the vertebrate embryo do not appear in invertebrate animals.

4. A vertebrate animal may be destitute of limbs; but when these are present, they never exceed four in number, are attached to an internal skeleton, and are turned *from* the neural side of the body. In the invertebrata the limbs are usually more numerous, are attached to a skeleton, which is external, and are turned *towards* the neural side of the body.

5. The jaws of vertebrate animals form part of the "parietes" of the head, and are never modified limbs, as in many invertebrata. Moreover, they move in a longitudinal and vertical plane; while the "foot-jaws" of the invertebrata, which possess them, meet horizontally, or else in a plane transverse to the body axis.

6. The blood system in vertebrata is always *vascular*, *i.e.*, enclosed in blood-vessels, with definite wall structures. All have a single valvular heart, except *amphioxus*. In invertebrate animals the blood, when present, often collects in the "perivisceral cavity," or in blood sinuses, hollowed out in the tissues.

7. Vertebrate animals have what is called the *hepatic portal system*, *i.e.*, the venous blood of the intestinal canal, after collecting in the *portal vein*, ramifies through the liver, and is exposed to the glandular action of that organ previous to its reception into the heart. No such arrangement is observed in invertebrata.

GLOSSARY.

The roots in the following are either Greek or Latin. The former are printed in the ordinary Greek characters; the latter in Italics.

Abdomen (*abdere*, to conceal), in vertebrates, the part of the trunk containing the intestines; in arthropoda, the hinder division of the body.

Acephala (α, privative; κεφαλή, the head), headless; applied to the lamellibranchiate molluscs.

Acetabulum (*acetum*, vinegar), a vinegar bowl; applied to the socket of the hip joint.

Acotyledones (α, privative; κοτυληδών, a cup-shaped hollow), applied to cryptogamic plants whose seed or "spores" have no cotyledons.

Acrogen, (ἄκρον, the highest point; γεννάω, to produce), a plant whose growth takes place at the summit, such as ferns, mosses, charæ, etc.

Actinia (ἀκτίν, a ray), applied to the sea-anemone, from its having the tentacles disposed radially. **Actinozoa** (ζῶον, an animal), a division of cœlenterata, whose type is the sea-anemone.

Adductor (*adducere*, to tighten), the muscle which closes the valves in bivalve molluscs.

Adnate (*ad*, to; *natus*, born), applied to stipules which adhere to the petiole.

Adult (*adolescere*, to grow to maturity), a full grown animal.

Aerial (*Aër*, air), existing in the air (= **sub-aerial**). To be **aerated** is to be exposed so as to absorb free oxygen from the air or water.

Agamogenesis (α, priv., γάμος, marriage; γένεσις, birth), applied to forms of reproduction, such as fission, etc., in which the sexes are not concerned (= asexual reproduction).

Alæ (*ala*, a wing), the lateral petals of a papilionaceous flower, such as that of the pea or bean.

Albumen (*albus*, white), an organic compound, such as white of egg, containing the four elements, carbon, oxygen, hydrogen, and nitrogen.

Albumen, the nutritious matter contained in most seeds, also called *perisperm*, and *endosperm*. **Albuminous** seeds are those which have an albumen; such as have no albumen are termed **ex-albuminous**.

Alburnum, the young wood of an exogenous stem.

Alimentation (*alimentum*, food), the process by which the food materials are prepared for absorption into the blood.

Ambulatory (*ambulare*, to walk), formed for walking; applied to the limbs, used for that purpose in the crustacea.

Amœba (ἀμοιβή, change), one of the protozoa, whose form is continually changing. **Amœbiform**, changing shape, like an amœba.

Amphibia (ἀμφί, both; βίος, life), animals which in the young state breathe by gills, but, in the adult condition, by lungs.

Anastomosis (ἀναστομοῦν, to open into), the opening of vessels or veins into each other, so as to form a kind of network.

Animalcule (*animalculum*), a very small animal.

Ankylosis (ἀγκύλος, crooked), the coalescence of two bones by which motion upon each other is prevented.

Anodon (ἀν, privative; ὀδούς, a tooth), toothless; a bivalve mollusc whose valves have no teeth at the hinge.

Annulus (a little ring), the elastic rim of the sporangium of fern.

Annulosa, in zoology, a sub-kingdom consisting of animals whose bodies are made up of a series of homologous *rings* or "somites" with their appendages.

Antebrachium (*ante*, before; *brachium*, the arm), the forearm of a vertebrate animal, formed of the radius and ulna.

Antenna (Lat. the sail yard of a ship), one of the "feelers" of insects and crustaceans.

Antennules, the smaller antennæ in crustaceans.

Anther (ἀνθηρός, blooming), the sac of the stamen containing the pollen.

Antheridium, the part which contains the **Antherozoids**, or male element in cryptogams.

Anura (ἀν, privative; οὐρά, a tail), tailless, applied to frogs, toads, and other tailless amphibia (=*Batrachia*.)

Aorta (ἀορτή, the large artery), an artery which arises directly from the heart.

Appendicular (*appendix*, an appendage), applied to the parts annexed to the main body of an organism, such as the limbs of an animal, the branches, etc., of a plant.

Arachnida (ἀράχνης, a spider; εἶδος, form), a class of anthropods of which the common spider is the type.

Arachnoid (ἀράχνη, a cobweb; εἶδος, form), like a cobweb; a very thin membrane covering the brain and spinal cord.

Archegonium (ἀρχή, beginning; γόνος, seed); the part containing the germ cells, or female elements in cryptogamia.

Arthropoda (ἄρθρον, a joint; πούς, a foot); the classes of annulosa which have jointed limbs.

Artery (ἀρτηρία), a vessel which conveys blood from the heart.

Asexual (=non-sexual), applied to modes of reproduction, such as gemmation, fission, etc., in which the sexes are not concerned.

Assimilation (*ad*, to; *similis*, like); the process by which the materials of the food are made to form part of the body substance.

Atlas ("Ατλας, the supporter of the heavens); the uppermost vertebra upon which the skull rests.

Atrophy (α, priv., τρίφω, to nourish); the wasting away of tissue, or part of the body substance for want of due nourishment.

Auricles (*auricula*, dim. of *auris*, the ear); the chambers of the heart which receive the blood from the veins.

Axilla (Lat. the arm-pit), the point of the angle formed by a leaf with the stem. **Axillary**, belonging to the axilla.

Axis (Lat. a pivot), the main line of a body or figure round which all the parts arrange themselves; applied to the trunk of a plant or animal as distinguished from the appendages; also to the second cervical vertebra. **Axile, axial**, pertaining to the axis.

Bacterium (βακτήριον, a staff), a rod-like jointed mobile filament, which appears in organic infusions when they begin to putrefy.

Basipodite (βάσις, a pedestal; πούς, a foot), in the arthropoda, that part of the limb which is jointed to the body.

Bast or **Bass**, the innermost bark of an exogen, consisting of fine woody fibre.

Batrachia (βάτραχος, a frog), applied by Huxley to the tailless amphibians (= *Anura*).

Bilateral Symmetry (*bis*, in two ways; *latera*, sides; συμμετρία, proportion). The similarity or correspondence of parts on the right and left sides of a body or figure.

Biology (βίος, life; λόγος, discourse), the science which treats of the nature and effects of life.

Bisexual (*bis*, in two ways; *sexus*, sex), applied to animals or plants that have not the sexes united in the same individual; opposed to *hermaphrodite*.

Bivalve (*bis*, twice; *valvæ*, folding doors), a mollusc having two shells, united by a hinge.

Blastema (βλάστημα, growth, increase), applied to a fluid capable of being at once differentiated into a definite tissue.

Blastoderm (βλαστός, a bud or sprout; δέρμα, skin), the thin membrane of the ovum, in which the embryo first appears.

Brachium (βραχίων, the arm), in vertebrates, the part of the fore-limb formed by the *humerus*.

Bract (*bractea*, a thin plate), the modified leaf commonly found at the base of a flower stalk.

Branchiæ (βράγχια, a gill), gills. **Branchial Chamber**, the cavity containing or contained by the gills. **Branchiostegite** (στέγω, I cover), a covering or protection for the gills.

Bronchi (βρόγχος, the windpipe), the two tubes into which the trachea divides, one leading to each lung. The smaller air passages are termed **Bronchial Tubes**.

Bulb (*bulbus*, an onion), an

underground bud, fleshy, and covered with scales.

Bulbus Arteriosus (βολβός, an onion), an enlargement at the root of an artery, usually contracting rythmically.

Byssus (βύσσος, fine flax), applied to the silky threads by which mussels and some other molluscs attach themselves to rocks, etc.

Cæcal (*cæcus*, blind), applied to a tube having one end closed, and hence said to terminate *blindly* or *cæcally*. **Cæcum**, a process of the large intestine which "ends blindly."

Calcify (*calx*, lime; *fieri*, to become), to become impregnated with carbonate or phosphate of lime.

Calibre, the wideness of a tube.

Calyx (κάλυξ, a flower cup), the *outer* whorl of leaves in a flower.

Cambium, the formative fluid in vegetables, yet undifferentiated, but which, like the blood in animals, is used in building up the tissues.

Canaliculi (*canaliculus*, a little channel), very minute channels uniting the "lacunæ" in bones with each other.

Capillary (*capillus*, hair) hair-like; applied especially to the very slender blood-vessels which unite the arteries with the veins.

Carapace, a shield or covering; that part of the exoskeleton that covers the cephalothorax in crustaceans.

Carina (*carina*, a keel), the two united lower petals of a papilionaceous flower.

Carotid (κάρα, the head; ὦτα, the ears), applied to the arteries of the head, which pass up close to the ears.

Carpel (καρπός, fruit) the modified leaf which goes to form the pistil, and contains the ovules. A pistil may consist of one or more carpels.

Carpus (καρπός, the wrist), the small bones between the forearm and the hand; the *wrist* in man; the *knee* in the horse.

Caudex (Lat.), the stem of palms and tree ferns.

Caulis (Lat. a stalk), an aërial stem. **Caulinary**, applied to stipules growing free of the petiole and of each other.

Centrum, (κέντρον, a point, a spur), applied to the "bodies" of the vertebræ.

Cephalic (κεφαλή, the head), belonging to the head. **Cephalo-thorax** (θώραξ, a breastplate, a cuirass), the part of a crustacean or arachnid, formed of the coalesced head and thorax.

Cerebrum (Lat.), the brain. **Cerebellum**, the *little brain*, the part overlying the fourth ventricle. **Cerebral**, belonging to the brain.

Cervical (*cervix*, the neck), belonging to the neck.

Chalaza (χάλαζα, hail, a pimple or tubercle), the point where the ovule of a flower is organically connected with the placenta; also the string by which the yolk of the egg is suspended.

Chelæ (χηλή, a claw), the "pincers" in crabs, lobsters, and their allies, *chelate*, possessing chelæ, applied to a limb,

Chitin (χιτών, a tunic), the horny substance forming the hard coating of insects, and others of the annulosa. It also forms the epidermic covering of encysted animalcules.

Chlorophyll (χλωρός, green; φίλλον, a leaf); the green colouring matter of leaves.

Chondrin (χόνδρος, cartilage), a substance obtained by boiling cartilage.

Chorda dorsalis (χορδή, a string; *dorsum*, the back), an embryonic structure in vertebrates, occupying the position afterwards taken by the bodies of the vertebræ (= *notochord*).

Chyle (χυλός, juice), the digested food in the condition in which it is absorbed by the lacteals into the blood circulation.

Chyme (χυμός, juice), the food after being dissolved by the gastric juice, in which condition it is *poured* out of the stomach into the intestines.

Cilia (*cilium*, an eye-lash), microscopic filaments projecting from cells or animalcules, and moving rythmically.

Clavicle (*clavis*, a key), the collar bone of the pectoral arch in vertebrates.

Cloaca (Lat. a sink), the cavity which receives the contents of the alimentary canal, as also of the genital and urinary ducts, in some invertebrate animals, and in monotremes, birds, reptiles, and amphibia.

Cnidæ (κνίδη, a nettle), the stinging "thread-cells" of the cœlenterata (= *nematocyst*).

Coagulate (*coagulare*, to curdle a fluid), to turn from a fluid into a concrete condition, as in the coagulation of the blood.

Coccyx (κόκκυξ, a cuckoo), applied to the terminal part of the spinal column in animals that have no proper tail. In man, it has the shape of a cuckoo's beak; hence the name.

Cochlea (κοχλίας, a snail's shell), applied to the spiral part of the labyrinth of the ear.

Cœlenterata (κοῖλος, hollow; ἔντερα, the entrails), the sub-kingdom which comprises the hydrozoa, and the actinozoa.

Coleorhiza (κολιός, a sheath; ῥίζα, a root), a sheath which encloses the radicle of an endogenous embryo.

Colloid (κόλλα, glue; εἶδος, form), a substance imperfectly soluble, and which in consequence cannot diffuse itself through an organic membrane.

Collum (Lat. the neck), the part where the plumule and radicle of an embryo plant unite.

Columella (dim. of *columna*, a pillar), the bone in the ear of some amphibia and other animals, which answers to the *stapes* in mammalia.

Column (*columna*), applied to the cylindrical body of the sea-anemone.

Commissure (*commissura*, a joining), applied to the nervous cords which unite different ganglia.

Condyle (κόνδυλος, a knuckle), the articulating surface of a bone; applied especially to the prominences on the skull by which it articulates with the atlas, or first vertebra.

Confervæ (*confervere*, to unite), vegetable organisms consisting of filaments, formed of a string of cells united end to end. **Confervoid**, jointed like a conferva.

Conjugation (*conjugare*, to unite), a peculiar form of vegetable gamogenesis.

Conidia (κόνις, dust), the minute spores of fungi (sing. *conidium*). **Conidiophore** (φορέω, to bear), the supporters of the conidia.

Connate (*con*, together; *natus*, born), applied to stipules that coalesce with each other.

Connective tissue (*connectere*, to unite together), a fibrous structure whose chief use is to unite the various parts of the body together.

Continuous gemmation (*continere*, to hold fast; *gemmare*, to bud), that form of gemmation in which the young beings, budded off, remain attached to the parent organism.

Coracoid (κόραξ, a crow; εἶδος, form), the second clavicle found in amphibia, reptiles, birds, and monotremes. In the higher mammals it is a mere process of the shoulder-blade, and shaped like a crow's beak; hence the name.

Coralligena (κοράλλιον, coral; γεννάω, to produce), an order of actinozoa which secrete the substance called "coral."

Cornea (*corneus*, horny), the transparent part of the outer coat of the eye.

Cornu (Lat. a horn; pl. *cornua*), a horn, projection.

Corolla (*coronella*, dim. of *corona*, a crown), the inner envelope of a flower.

Corpora bigemina (*corpus*, a body; *bigemina*, twins), two bodies that overlie the passage from the fourth to the third ventricle in fishes, amphibia, reptiles, and birds. In mammals, a transverse depression divides these "optic lobes," as they are also termed, into *four bodies*, hence called **Corpora quadrigemina**.

Corpora striata (*stria*, the fluting of a column), "striped bodies," of which one is found in each of the lateral ventricles of the brain.

Corpus callosum (*callosus*, having a hard skin), the great band of nerve matter which unites the two hemispheres of the brain in mammals.

Corpuscle (*corpusculum*, a little body), a very minute cell. **Corpusculated**, applied to a fluid, such as the blood, containing corpuscles.

Cortex (Lat. bark). **Cortical**, belonging to the bark, applied to the outer part of an organ, as the cortical layer of the kidneys.

Costa (Lat.), a rib. **Costal**, pertaining to the ribs, as costal respiration.

Cotyledon (κοτυληδών, a cup-shaped hollow), the temporary leaf of an embryo plant.

Cranium (Lat.), the skull. **Cranial**, belonging to the skull.

Craspedum (κράσπεδον, a border or edge), applied to the convoluted cords attached to the free edges of the mesenteries in the actinozoa.

Crus (Lat. the shin), the part of the hinder limb of a vertebrate, consisting of the tibia and fibula.

Crustacea (*crusta*, a crust or shell); a class of arthropods, like the lobster, crab, etc., possessing a hard exoskeleton, which they shed periodically.

Cryptogamia (κρυπτός, hidden; γάμος, marriage), plants in which the reproductive process is hidden or obscure—not by flowers and seeds as in the *phanerogamia*.

Ctenophora (κτείς, a comb; φερέω, to bear), an order of actinozoa which swim by means of cilia attached to plates formed like a comb.

Culm (*culmus*), the stem or stalk of grasses.

Cutis (Lat.), the deeper part of the skin which contains nerves and blood-vessels, also called *corium* or *derma*. **Cuticle** (*cuticula*), the superficial part destitute of nerves and blood-vessels (= *epidermis*). In plants, the cuticle is a thin layer covering what is called the "epidermis."

Cyclosis (κυκλόω, to whirl round), a circulation of the fluid contents of a cell.

Cyst (κύστις, a bladder), a sac, an investing membrane.

Dehiscence (*dehiscere*, to gape, to split), the bursting of an envelope, such as an anther or seed-vessel when the contents are ripe.

Derma (δέρμα, the skin), applied to the *cutis vera*, or vascular layer of the integument.

Diagram (διάγραμμα), a figure intended to show the relation of certain parts of an object to each other—not given as an exact representation of what occurs in nature.

Diaphragm (διάφραγμα, a partition, the midriff, also called φρήν, whence the adj. *phrenic*), the muscle which in mammals separates the thorax from the abdomen.

Diastole (διαστέλλω, to separate or expand), the dilatation or expansion of one of the cavities of the heart immediately after its contraction or *systole*.

Diatom (διατέμνω, to cut through), a minute vegetable organism, provided with a flinty covering.

Dicotyledone (δίς, double; κοτυληδών, a cup-shaped hollow), a plant whose embryo has two cotyledons, or temporary leaves.

Differentiation (*differre*, to separate), the process by which cell structures are transformed into the various tissues of a plant or animal.

Digit (*digitus*), a finger or toe.

Disc (δίσκος, a quoit, a circular plate), a flat circular surface; applied to the space round the mouth of an actinia.

Discontinuous gemmation, that form of gemmation in which the young organisms, budded off, separate from the parent, and lead an independent existence.

Disk, a part intervening between the stamens and pistil of a flower.

Dorsal (*dorsum*, the back), pertaining to the back—opposed to *ventral*.

Duct (*ductus*, a passage for water), one of the vessels of plants; the outlet of a gland.

Duodenum (*duodeni*, by twelves), the part of the intestine immediately connected with the stomach; so named from its supposed length in man.

Dura mater (Lat. "rigorous mother"), the tough fibrous membrane that lines the cavity of the skull and the neural canal.

Duramen (Lat. hardness, durability), the "heart wood" of exogenous trees.

Ectoderm (ἐκτός, outside; δέρμα, skin), the outer integumentary layer in the cœlenterata.

Embryo (ἔμβρυον, from ἐν, in, and βρύω, to swell), the earliest stage in which the young animal appears in the ovum; the young plant as it is contained within the seed. **Embryogeny** (γεννάω, to produce), the production of the embryo.

Emulsion (*emulgere*, to milk out), the milky matter contained in the chyle, formed by the action of the bile and pancreatic juice upon the fatty matters of the food.

Encephalon (ἐν, within; κεφαλή, the head); the brain, inclusive of the *medulla oblongata*.

Encysted (ἐν, within; κύστις, a bag); the condition of some among the protozoa (the amœba, for instance), when they become motionless, and surround themselves with a hard coating.

Endoderm (ἔνδον, within; δέρμα, the skin); the inner cellular layer in the cœlenterata.

Endogen (γεννάω, to produce), a plant wherein the production of new matter is internal. **Endogenous**, growing internally. **Endogenous cell multiplication**, a mode of cell development in which the *contents* of the original cell break up into a number of distinct masses, each of which becomes a separate cell. The "cleavage of the yelk" in the fecundated ovum is a good example.

Endolymph (*lympha*, clear water), the fluid within the semicircular canals of the inner ear, and containing the *otoliths*.

Endophlœum (φλοιός, bark), the liber, or inner bark of a tree.

Endophragma (φράγμα, a protection), an ingrowth of chitin in the lobster intended to form a roof for the nerve chamber in the under part of the thorax.

Endopodite (πούς, a foot), "the internal distal segment of the typical limb of crustacea."

Endorhizal (ῥίζα, a root), applied to the germination of monocotyledones, in which the rootlets spring forth from *within* the embryo.

Endoskeleton (σκελετός, dried up), the internal *hard* parts of an animal, such as the bones and cartilages.

Endosmosis (ὠθέω, to push, to impel), the movements of fluids *inwards* through a membrane.

Endosperm (σπέρμα, seed), the

store of nutriment (albumen) laid up in the seed for the temporary support of the embryo. Applied also to the inner of the two coats of a fern spore.

Epiblast (ἐπί, upon; βλαστός, a bud), the outer layer of the germinal membrane in the impregnated ovum.

Epiblema (ἐπίβλημα, a covering), the integument of the root of a plant.

Epidermis (ἐπί, upon; δέρμα, the skin), the outer, cellular, or horny layer of the integument in animals; the external cellular covering of the bark and leaves of plants.

Epibranchial (βράγχια, a gill), applied to the *internal* cavities formed by the gills in lamellibranchiate bivalves.

Epiphlœum (φλοιός, bark), the *corky*, or outer cellular layer of bark in exogenous trees.

Epipodite (πούς, a foot), a process from the basal joint of some of the limbs of certain crustacea.

Epithelium (θῆλυς, delicate, tender), a cellular layer lining the internal cavities, tubes, and vessels of the body.

Exalbuminous (*ex*, without; *albumen*), applied to seeds destitute of an albumen.

Excretory (*excernere*, to separate, to sift), applied to those organs in a plant or animal which separate and remove the products of decay.

Exhalation (*exhalare*, to breathe out), the giving off of gas or vapour into the atmosphere.

Exogen (ἔξω, outside; γεννάω, to produce), a plant in which the growth is external. **Exogenous**, growing externally.

Exopodite (ἔξω, outside; πούς, a foot), "the external distal segment of the typical limb of crustacea."

Exorhizal (ῥίζα, a root), applied to the germination of dicotyledones, in which the radicle is a direct continuation of the axis, and external to the rest of the embryo.

Exoskeleton (σκελετός, dried up), the external hard part of certain animals, such as the shells of molluscs, the crust of crustaceans, etc.

Exosmosis (ὠθέω, to push), the movements of fluids *outwards* through a membrane.

Exosperm (σπέρμα, seed), the outer coating of a fern spore.

Extine (*ex*, outwards), the external coating of a pollen grain.

Fascia (Lat. a bandage), the envelope which encloses a bundle of striped muscular fibres.

Fascicle (*fasciculus*, a little bundle), applied to the collection of rootlets, all growing from one point, in endogens. **Fasciculate**, arranged in bundles.

Fecundation (*fecundus*, fruitful), the action of the sperm cell upon the germ cell, by which the latter is rendered capable of producing an embryo.

Femur (Lat.), the thigh bone.

Fermentation (*fermentum*, leaven), the chemical action set up in organic matter, by which its nature is changed, and certain gases are evolved.

Fertilisation (*fertilis*, fruitful), =*fecundation*, *q.v.*

Fibrilla (dim. of *fibra*, a thread), one of the minute threads into which striped muscular fibre may be resolved.

Fibrin (*fibra*, a thread), an element in the blood, which on coagulation assumes the form of very fine fibres.

Fibula (Lat. a pin or brooch), the slender bone which accompanies the tibia of the *crus* in vertebrates.

Filament (*filum*, a thread), any structure of thread-like shape; the stalk which supports the anther.

Fission (*fissio*, a splitting or cleaving), asexual reproduction, in which young organisms are produced by spontaneous division of the parent.

Follicle (*folliculus*, a little bag), a pit, or depression in the surface of a mucous membrane, in which a peculiar fluid is secreted; one of the cavities of a gland.

Foramen (Lat.), an aperture. The part of the ovule left uncovered by its coats (=*micropyle*). **Foramen magnum**, the opening in the base of the skull through which the cerebro-spinal axis passes.

Foraminifera (*ferre*, to bear), an order of protozoa having their shells pierced by numerous minute openings.

Formative material (*formare*, to form; *materia*, the stuff of which a thing is made), matter already fitted to become part of a living organism.

Fovilla (Lat. =*favilla*, dust or ashes), the minute granular matter contained within a pollen grain.

Fronds (*frondes*, chaplets of leaves), the *leaves* of ferns.

Function (*fungi*, to perform one's duty), the work which an organ is fitted to perform.

Funiculus (dim. of *funis*, a rope), the little cord by which the ovule is attached to the placenta of the carpel.

Gamogenesis (γάμος, marriage; γίνεσις, origin), reproduction by the mingling of the protoplasms of two distinct cells (=*sexual reproduction*).

Ganglion (γάγγλιον, a lump or swelling), a knot of nerve matter containing nerve cells or **ganglionic corpuscles**, and giving origin to nerve fibres.

Gastric (γαστήρ, the belly, the stomach), relating to the stomach, as the "gastric follicles." **Gastric juice**, an acid fluid secreted by the peptic glands, which abound in the walls of the stomach; the fluid dissolves all the *nitrogenous* parts of the food, turning them into *peptone*.

Gelatine (*gelare*, to freeze), a substance obtained by boiling connective tissue, and which turns into a *jelly* on cooling.

Gemmation (*gemma*, a bud), reproduction by *budding*; **continuous**, when the young organisms remain united with the parent; **discontinuous**, when they drop off and live independently.

Gemmules (dim. of *gemma*), the ciliated embryos of some coelenterata; also the encysted masses of sponge particles from which new ones are produced.

Germ cell (*germen*, a bud), the cell which, after receiving the contents also of the "sperm cell," produces the embryo.

Germinal cell or **vesicle**, the part of the ovule which, in fecundation, receives the contents of the pollen grain; also applied to a cell in the unimpregnated ovum, which contains in its wall a nucleus called the **germinal spot**. **Germinal area**, the part of the blastoderm from which the embryo is produced.

Germination (*germinare*, to bud), the first sproutings of the embryo from the seed.

Gland (*glans*, an acorn), applied to an organ in animals which "secretes" certain substances from the blood, which are afterwards passed out by a duct. The organs which produce peculiar secretions in plants are also called "glands."

Glenoid (γλήνη, the pupil of the eye, a shallow depression: εἶδος, form), applied to the hollow in the scapula which receives the head of the humerus.

Glochidia (γλωχίς, the barb of a hook or arrow), applied to young molluscs as they are found within the gill cavities of the parent.

Glottis (γλῶττα, the tongue), the slit in the larynx, between the vocal cords, opening into the trachea. During the act of swallowing, it is covered by an elastic tongue-shaped gristle, called the **epi-glottis**.

Hæmal (αἷμα, blood), pertaining to the blood. In invertebrate animals, the **hæmal aspect** corresponds with the "dorsal side" of the body, as opposed to the *neural aspect*, or ventral side.

Hallux (*hallex*), the great toe; the innermost of the five digits of the foot.

Hermaphrodite (ἑρμαφρόδιτος), an animal that unites both sexes in the same individual; applied also to flowers having both stamens and pistils.

Hilum (Lat. the black point of a bean), the part by which the seed is attached to the funiculus.

Histology (ἱστός, a web; λόγος, a discourse), the minute structure of the tissues.

Homogeneous (ὁμός, alike; γένος, kind), having the same texture and consistence throughout.

Homology (ὁμολογία, agreement), the relation of parts that are structurally similar, though they may differ a good deal in function, as the *arm* of a man, the *fore leg* of a horse, and the *wing* of a bird.

Humerus (Lat.), the bone of the upper arm (*brachium*) in vertebrates.

Humus (Lat.), soil formed of decomposed organic substances.

Hydra (ὕδρα, a water dragon), the common fresh-water polype, the type of the class **Hydrozoa** (ζῶον, an animal).

Hymenium (ὑμήν, a membrane),

15B M

the part of a mushroom which is formed of a number of lamellæ, or radiating partitions.

Hyoid (the letter υ; εἶδος, form), U-shaped; applied to the bone at the upper part of the windpipe.

Hypha (ὑφή, a weaving), one of the filaments which go to form the mycelium of a fungus.

Ilium (εἰλίω, to shut in), the upper part of the *os innominatum*.

Imbibition (*imbibere*, to drink in), the absorption by organic structures of the fluids surrounding them.

Imbricated (*imbrex*, a roof tile), arranged with one part overlapping the next, like the tiles or slates on a roof.

Indusium (*induere*, to clothe), an epidermic covering of the sporangia in some ferns.

Infundibulum (Lat. a funnel), a hollow tube connecting the pituitary body with the third ventricle of the brain.

Infusoria (*infusum*, from *in*, into, and *fundere*, to pour), a class of animalcules so named from their occurrence in organic infusions.

Innominate bone (= *os innominatum*, a bone without a name), each of the two bones which, with the *os sacrum*, make up the pelvic arch.

Insertion (*insertio*, an engrafting upon), the point by which a muscle is attached to the part which is to be moved.

Inspissation (*spissare*, to thicken, condense), the thickening of a vegetable fluid or juice by the evaporation of moisture from the leaves.

Integument (*integumentum*, a covering), the external protective covering of a plant or animal.

Intercellular (*inter*, between; *cella*, a cell), applied to the spaces between or among cells.

Internodes (*nodus*, a knot), the spaces between the nodes on the stem of a plant.

Intervertebral (*inter*, between; *vertebra*, a joint of the back), applied to the cartilages, apertures, etc., occurring between two vertebræ.

Intestines (*intestinum*, a gut), the portion of the alimentary canal below or behind the stomach.

Intine (*in*, within), the inner coating of a pollen grain.

Intussusception (*intus*, within; *suscipere*, to take up), the act of taking foreign matter into a living body.

Iris (ἶρις, a rainbow), the coloured part of the eye which surrounds the pupil.

Ischium (ἰσχίον, the hip), one of the pelvic bones. **Ischiatic** (= *sciatic*), pertaining to the ischium.

Labium (Lat. a lip), the under part of the mouth in arthropoda (= *metastoma*).

Labrum (Lat. the upper lip), the upper part of the mouth in arthropoda.

Lacteals (*lac*, milk), the vessels of the intestinal villi which absorb the milk-like chyle.

Lacunæ (Lat. cavities), interspaces among the tissues of plants or animals; applied especially to the little elliptical hollows found in bone.

Lamellæ (dim. of *lamina*, a thin plate), applied to the thin plates in the hymenium of a mushroom; also to the gills of **lamellibranchiate** molluscs.

Lamina (Lat. a thin plate), applied to the expanded part or blade of a leaf or frond.

Laminated, disposed like the leaves of a book.

Larva (Lat. a mask), applied to the earliest active forms of insects, and of some molluscs and amphibia, which differ considerably in appearance from the adult animals—*e.g.*, caterpillars, zoeæ, and tadpoles.

Leptothrix (λιπτός, slender; θρίξ, hair), a fine filamentous body accompanying the germination of the bacterium.

Leucocythæmia (λευκός, clear; κύτος, a cell; αἷμα, blood), a disease in which the blood is largely charged with colourless corpuscles.

Liber (Lat.), the inner bark (= *endophlœum*).

Ligament (*ligamentum*, a band), a membrane or muscle uniting two other structures, usually bones; applied to the "hinge" of bivalves.

Lignine (*lignum*, wood), the essential matter of woody fibre, formed of hardened cellulose.

Limb, applied to the expanded part of a petal.

Lobe (λοβός, the lower part of the ear, the *lobe* of the liver), applied to parts similarly shaped or divided, as the lobe of a leaf, etc.

Locomotion (*locus*, a place; *motio*, motion), change of place.

Longitudinal (*longitudo*, length), from end to end; applied to a "section" extending from one end of an object to the other.

Lymph (*lympha*, clear water), the blood *serum* which flows through a system of vessels called the **lymphatics**.

Mammalia (*mammæ*, paps), the class of vertebrates which suckle their young.

Mandible (*mandibulum*, from *mandere*, to chew), the lower jaw in vertebrate animals; in the arthropoda, the upper pair of cephalic appendages used as jaws; in birds, sometimes applied to both rostra of the beak.

Mantle, the *pallium* of a mollusc (see *pallium*). **Mantle cavity**, the hollow enclosed by the mantle.

Manubrium (Lat. a handle), a part of the segmented antheridium of chara, to which several jointed filaments are attached, like a whip with a number of thongs.

Manus (Lat.), the hand; the fore foot of a quadruped.

Masticatory (μασάομαι, to chew), applied to organs which are used for grinding the food, such as teeth, jaws, maxillipedes, etc.

Matrix (Lat. a parent stock), the substance in which things are embedded; *e.g.*, the semitransparent matter which surrounds cartilage cells.

Maxillipedes (*maxilla*, a jaw; *pes*, a foot), the *foot-jaws* of crustacea.

Medulla (Lat.), the pith of plants, or marrow of bones. **Medulla oblongata**, the part of the brain which passes into the spinal cord. **Medullary rays** or **plates**, thin sheets of cellular tissue uniting the pith with the cambium zone. **Medullary sheath**, the envelope of the pith, consisting largely of spiral vessels.

Membrana nictitans (Lat. the winking membrane), a third eye-lid, found in birds, amphibia, and in some mammals.

Mesenteries (μίσος, the middle; ἔντερα, the bowels), the vertical radial partitions which divide the somatic cavity of the actinia into chambers. **Mesentery**, the membrane which, in the vertebrata, supports the alimentary canal.

Mesophlœum (φλοιός, bark), the middle layer of the bark.

Metacarpus (μετά, next after; καρπός, the wrist), the bones of the *manus*, which give attachment to the digits.

Metamorphosis (μεταμόρφωσις, a change of form), applied to transformation of cell structures into the different tissues or secretions of an organism; applied also to the changes of form which some animals undergo from the larval to the adult condition.

Metatarsus (ταρσός, the flat of the foot), the bones of the foot, which support the toes.

Micropyle (μικρός, little; πύλη, a gate or entrance), the aperture in the coatings of the ovule through which the pollen tube passes to impregnate the germinal cell (= *foramen*).

Mobile (*mobilis*, movable, from *movere*, to move), capable of self-movement.

Mollusca (*mollis*, soft), a sub-kingdom, including what are called shell-fish, with several others, so named from the softness of their bodies.

Monocotyledone (μόνος, single; κοτυληδών, a hollow like a cup), a plant whose embryo has but a single cotyledon, or temporary leaf.

Monotremata (μόνος, single; τρῆμα, an aperture), an order of mammals which have but one exit, in the form of a cloaca, for the intestinal, genital, and urinal canals.

Morphology (μορφή, form; λόγος, discourse), the study of the form and structure of organised beings.

Motor nerve (*movere*, to move), a nerve which causes a muscle to contract.

Muscle (*musculus*), a structure in an animal body which contracts in obedience to the will, or on the application of the proper stimuli, and so gives rise to the various movements of which the body is capable.

Mycelium (μύκης, a fungus, a mushroom), a sort of loose felt-work, formed by the interlacing of hyphæ.

Myelon (μυελός, marrow), the spinal cord in vertebrate animals. **Myelencephalon** (ἐγκέφαλος, the brain), the *medulla oblongata*.

Myriapoda (μυρίος, countless; πούς, a foot), a class of arthropoda, embracing centipedes, millipedes, etc., and characterised by having numerous feet.

Nacre (an oriental word), mother of pearl; **nacreous**, pearly.

Nematocyst (νῆμα, a thread; κύστις, a bag), a "thread cell."

Nerve (*nervus*, a sinew, a nerve), applied to the veins of leaves; also to those tissues in animals by which motor or sensory impulses are conveyed. **Nervation**, the *venation* of leaves; in animals, the arrangement of the nervous system.

Neural (νεῦρον, a cord, a nerve), pertaining to the nervous system. **Neurilemma** (λίμμα, rind), the sheath which encloses a bundle of nerve fibres.

Nictitating membrane. [See *membrana nictitans*.]

Node (*nodus*, a knot), the part of a stem from which leaves or leaf-buds proceed.

Notochord (νῶτος, the back; χορδή, a string), a structure found in vertebrate embryos, forming the foundation of the subsequent *centra* of the vertebræ (= *chorda dorsalis*).

Nucleus (Lat. a kernel), a speck of germinal matter, usually found in the interior of cells; applied also to the central part of the ovule. **Nucleated**, possessing a nucleus.

Nucule (*nucula*, dim. of *nux*, a nut), a term applied to the "spore-fruit," or female element, in the fructification of chara.

Occipital (*occiput*, the back part of the head), pertaining to the back part of the skull.

Œsophagus (οἶσος, a reed; φαγεῖν, to eat), the gullet or tube leading from the mouth to the stomach. (*Vide Arist. De Part. An.* iii. 3).

Olfactory (*olfactus*, the sense of smell, from *olfacere*, to smell at), applied to the nervous structures which give rise to the sensation of smell.

Oogonium (ᾠόν, an egg; γόνος, seed), the germ cell of some fungi, which, on impregnation, becomes an **oospore** (σπορά, that which is sown; a *spore*).

Operculum (Lat. a lid, from *operire*, to cover), a little covering.

Optic (ὄπτομαι, to see), pertaining to vision. The **optic lobes** are the parts of the brain in vertebrata to which the **optic nerves** lead from the eye. In the lower classes of these animals there are two optic lobes, one on each side, hence called *corpora bigemina*. In the higher classes there are two on each side, four in all, and hence termed *corpora quadrigemina*.

Origin (*origo*, beginning, source), applied to that end of a muscle which may be regarded as fixed during contraction; the other end, which is thus drawn up nearer to the origin, is termed the *insertion*.

Organ (ὄργανον, an instrument), a part of a plant or animal to which some function is assigned. **Organism**, a being possessed of an organic structure; any plant or animal.

Otoconia (οὖς, the ear; κονία, sand or dust), minute sandy particles found in the vestibule of the ear; when they strike against the auditory nerve, the sensation of sound is produced.

Otoliths (λίθος, a stone), small calcareous concretions found in the ears of fishes, crustacea, and mollusca, which act like the otoconia in the human ear.

Ovum (Lat.), an egg, applied to the germ cells of animals; sometimes they are called **ovules**, if unimpregnated. The latter term is also applied to the immatured seeds of plants. **Ovary**, the special organ, by, or in which, the ovules are produced.

Oxygen (ὀξύς, sharp, acid; γεννάω, to produce), the principle which produces acidity. **Oxygenate**, to charge with oxygen. **Oxidation**, a chemical union with oxygen.

Pallium (Lat. a cloak), the "mantle" of mollusca. **Pallial impression**, a groove in a bivalve shell, marking the line by which it was attached to the margin of the mantle.

Palpi (*palpare*, to touch), supposed organs of touch in arthropoda and lamellibranchiata. In the former they are jointed limbs attached to the mandibles; in the latter they are little membranous appendages suspended from the sides of the mouth.

Pancreas (πᾶν, all; κρέας, flesh), a gland situated under the stomach; it secretes the **pancreatic juice**, which is poured along with the bile into the duodenum.

Papilionaceous (*papilio*, a butterfly), applied to flowers which, like those of the bean, pea, etc., have the corolla shaped somewhat like the wings of a butterfly.

Papilla (Lat. a nipple or teat), a minute soft prominence. **Papillated** and **papillose**, covered with minute outgrowths like nipples.

Parasite (παράσιτος, a toady), applied to a plant or animal which derives its nourishment from the living tissues or fluids of another plant or animal.

Parenchyma (παρά, together; ἐγχέω, to pour in), the cellular tissue of plants.

Parietes (Lat. walls), the walls of a cavity. In plants, the *placenta* is **parietal** when it is attached to the walls of the ovary. The **parietal** bones of the skull are those that form the middle part of its roof. **Parieto-splanchnic** (σπλάγχνα, the intestines), a pair of ganglia in molluscs, which supply filaments to the mantle, gills, and viscera.

Patella (Lat. a little dish), the knee-pan; applied also to the limpet.

Pectoral (*pectus*, the breastbone), pertaining to the

breast. Pectoral arch, the group of bones which support the fore limbs in vertebrates.

Pedal (*pes*, the foot), pertaining to the foot; applied to a pair of ganglia situated in the *foot* of molluscs.

Peduncle (*pediculus*, dim. of *pes*, a stalk), the stalk or axis of a flower or fern frond; the "eye-stalk" in the lobster, etc.

Pelvis (Lat. a basin), the bony arch which supports the hinder limbs of vertebrates.

Penicillium (*penicillum*, a painter's brush), the ordinary "green mould," a fungus so named from the brush-like appearance of its aërial hyphæ.

Perianth (περί, around; ἄνθος, a flower), applied to the "floral envelopes," *i.e.*, the calyx and corolla of a flower.

Pericardium (καρδία, the heart), the membrane, with its cavity, within which the heart is contained. In crustacea, this is simply a cavity hollowed out of the tissues, and forming a venous sinus.

Perilymph (*lympha*, clear water), the fluid which surrounds the labyrinth of the ear.

Periosteum (ὀστίον, a bone), the vascular membrane which invests the bones, and supplies them with blood.

Peripheral (περιφέρω, to encircle), applied to the outer or marginal parts of anything.

Perisperm (σπέρμα, a germ), a term sometimes applied to the *albumen* of the seed, more commonly called *endosperm*.

Peristaltic (στέλλω, to contract), applied to the movements of the intestines, in which, when a particular part of the canal contracts, the constriction moves along it like a wave in the water. It is also termed the *vermicular* movement (*vermis*, a worm), from its resemblance to the mode in which the rings of that animal are successively contracted during locomotion.

Peritoneum (τείνω, to stretch), the membrane which covers the abdominal walls and the contained viscera.

Perivisceral (*viscera*, the bowels), applied to the space surrounding the organs of digestion.

Petal (πέταλον, a leaf), a leaf of the corolla of a flower.

Petiole (*petiolus*, a stalk), a leaf stalk.

Phalanges (φάλαγξ, a row, also a round piece of wood), the small bones composing the digits of the higher vertebrates.

Phanerogam (φανερός, visible; γάμος, marriage), a plant whose fructification is conspicuous. The term includes all exogens and endogens.

Phyllodium (φύλλον, a leaf; εἶδος, form), a broad flattened petiole, resembling the blade of a leaf.

Phrenic (φρήν, the midriff), pertaining to the diaphragm; the movements of this organ are directed by the *phrenic nerve*.

Physiology (φύσις, nature; λόγος,

discourse), the study of the vital operations which take place in plants or animals.

Pia mater (Lat. tender mother), a delicate vascular membrane which invests the brain and spinal cord in vertebrates.

Pileorhiza ($\pi\tilde{\imath}\lambda os$, *pileus*, a cap; $\acute{\rho}\acute{\imath}\zeta a$, a root), the sheath protecting the growing cells of a young root.

Pileus (Lat. a cap), the cap-like part of a mushroom, which bears the hymenium or "gills" on its under side.

Pineal gland (*pinea*, a fir cone), a body shaped somewhat like a fir cone (and hence also called *conarium*), projecting above the roof of the third ventricle of the brain.

Pinnules (*pinnula*, a little wing or feather), the little lobes or leaflets of a fern frond, or of any other leaf similarly divided.

Pistil (*pistillum*, a pestle), the female organ of a flower, composed of one or more carpels; each carpel has usually an *ovary*, *style*, and *stigma*.

Pitchers (or *ascidia*, from $\dot{a}\sigma\varkappa\acute{o}s$, a leathern bottle), the leaves of certain plants hollowed into the form of pitchers, and containing a fluid the evaporation of which is prevented by a lid.

Pituitary body (*pituita*, phlegm), a body that depends from the brain just under the third ventricle, with which it is connected by the *infundibulum*.

Placenta (Lat. a cake), in the higher mammalia, a highly vascular organ by which the blood of the parent is applied to the nourishment of the *fœtus* (or young animal yet unborn); in plants, the part of the carpel to which the ovules are attached by the *funiculus*, and through which they derive their nourishment.

Pleuron ($\pi\lambda\epsilon\nu\varrho\acute{o}\nu$, a rib), applied to each of the lateral projecting parts in the somite of a lobster. **Pleura** ($\tau\grave{a}$ $\pi\lambda\epsilon\nu\varrho\acute{a}$, the ribs, the *side*), the membrane which covers the lungs, and the walls of the chest in air-breathing vertebrates.

Plexus (*plectere*, to entwine), an interlacing of nerve trunks or twigs.

Plumule (*plumula*, a little feather), the first sprout of the embryo plant.

Podophthalmata ($\pi o\acute{v}s$, a foot; $\dot{o}\varphi\theta a\lambda\mu\acute{o}s$, an eye), a division of crustaceans in which the eyes are supported on long foot stalks.

Pollen (Lat. fine flour), the fine powdery matter contained within the anther, and which is necessary for the fertilization of the ovules. **Pollen tube**, a long process from a pollen grain, which extends itself down through the style till it touches an ovule.

Polype ($\pi o\lambda\acute{v}\pi o\nu s$, many footed), applied to the hydra and actinia; and generally to the individual organisms (zoöids) of the compound actinozoa. **Polypary**, the chitinous covering of compound hydrozoa.

Portal vein (*vena portæ*, the vein of the gate), a vein which, after collecting the blood of the intestines, ramifies through the liver. A **portal system** refers to the redistribution of blood by a vein to a second set of capillaries, such as takes place in the liver and kidneys.

Posterior fissure (*fissura*, a cleft, from *findere*, to cleave), the longitudinal depression on the dorsal side of the spinal cord.

Prehensile (*prehendere*, to seize), having the power of grasping or laying hold of.

Primine (*primus*, first), the outer coat of the vegetable ovule.

Primordial utricle (*primordius*, original; *utriculus*, a little bag), the outer layer of the protoplasm of a cell.

Primitive groove (*primitivus*, original), a depression on the blastoderm of a fertilised ovum, in which the neural axis subsequently becomes developed.

Procœlous (πρό, before; κοῖλος, hollow) applied to vertebræ whose centra are hollow in front and convex behind.

Process (*processus*, a projection), a projection or outgrowth from any part of an organism (generally pronounced *prō-cess*).

Proembryo (πρό, previous to; ἔμβρυον, embryo), a chain of cells which grow from the spore of a chara, and from which arises the embryo.

Pronation (*pronus*, facing downwards, stooping), the turning of the hand with the palm downwards; opposed to *supination*.

Protein (πρωτεύω, to have the first place), an organic compound consisting of carbon, hydrogen, oxygen, and nitrogen, and supposed (erroneously) by Mulder to be the basis of all the albuminoids.

Proteid (εἶδος, resemblance), any substance resembling or containing protein.

Prothallium (πρό, previous to; θαλλός, a young shoot), the immediate result of the germination of a fern spore; its function is to engender the embryo plant.

Prosenchyma (πρός, beside; ἐγχέω, to pour in), a name given to the woody fibre in plants.

Protococcus (πρῶτος, first; κόκκος, a kernel or berry), a vegetable organism consisting of but one cell.

Protoplasm (πλάσμα, from πλάσσω, to shape or fashion), the living active matter in a cell, which is constantly absorbing new material from without, and manufacturing *tissues* out of its own substance, the change being called *differentiation*.

Protopodite (πούς, a foot), the basal part of the typical limb of a crustacean.

Protovertebræ (*vertebra*, a joint of the back), rings which, in the embryo vertebrate, surround the notochord, and become the bases of the future vertebræ.

Protozoon (ζῶον, an animal), the lowest division of the animal kingdom.

Protractor muscle (*protrahere*, to draw forward; *musculus*, a muscle), a muscle in the foot of the anodon by which it is moved forward.

Pseudopodia (ψευδής, false; πόδις, feet), projections thrust out from the body substance of a rhizopod, which can also be fully retracted. They serve the purpose of limbs both for prehension and locomotion.

Pulmonary (*pulmones*, the lungs), pertaining to the lungs. **Pulmonate**, possessing lungs. **Pulmo-cutaneous** (*cutis*, the skin), pertaining to the lungs and skin, applied to arteries in the frog which supply blood both to the lungs and skin.

Pupil (Lat. *pupilla*), the aperture in the iris which affords passage to the rays of light entering the eye.

Pylorus (πύλη, a gate; οὖρος, a guard), the valve guarding the passage leading from the stomach to the intestine.

Quaternary (*quaterni*, four together), consisting of four, or some multiple of that number; applied to the sepals and petals of some exogens.

Quiescent (*quiescere*, to repose), applied to that form of the protococcus which is incapable of self-movement; opposed to *mobile*.

Quinary (*quini*, five together), consisting of five, or some multiple of five; applied to the number of parts in some exogenous flowers.

Radicle (dim. from *radix*, a root), the root of an embryo plant.

Radius (Lat. a spoke), the inner of the two bones of the fore arm, the other being called the *ulna*. **Radial**, diverging like rays from a central point, as the "mesenteries" and tentacles of the actinia.

Ramus (Lat. a branch), a side or half of the mandible or lower jaw of vertebrates.

Rectus (Lat. straight), applied to the muscles of the eye which move it up or down, or from side to side. **Rectum**, the terminal part of the intestine.

Renal (*renes*, the kidneys), pertaining to the kidneys.

Respiration (*respirare*, to breathe), the inhaling and exhaling of air either by plants or animals.

Restiform (*restis*, a rope), a tract of fibres in the medulla oblongata which support the cerebellum.

Reticulated (*rete*, a net) disposed like the threads in a net, as the *veins* of leaves, the *capillaries* of the blood-vascular system, and the nerve fibres in the **retina** of the eye.

Retractor (*retrahere*, to draw back), applied to the principal pair of muscles in the "foot" of an anodon.

Rhizoid (ῥίζα, a root; εἶδος, resemblance), one of the rootlets in chara. **Rhizome** (ὁμός, the same as), the underground stem of ferns, and some other plants.

Rigor mortis (Lat. the stiffness of death), a term applied to the rigidity of the muscles soon after death.

Rostrum (Lat. the beak of a bird), applied to the part which projects from the head of a lobster or other crustacean.

Rotifera (*rota*, a wheel; *ferre*, to bear), animalcules provided with cilia round the mouth, the movements of which resemble, in appearance, the revolution of toothed wheels.

Rythmical (ῥυθμός, measured time), acting at regular intervals, as the pulsations of the heart.

Sac (σάκκος, a bag), any pouch-shaped organ; the wall of a cell.

Saccharine (σάκχαρον, sugar), consisting of, or containing sugar.

Sacrum (*sacrum*, sc. *os*, the sacred bone), the vertebræ, to which the innominate bones of the pelvis are attached. **Sacral**, pertaining to the *os sacrum*.

Sap, the juice of a plant. **Sapwood**, the *alburnum* or young wood through which the crude sap ascends.

Sarcode (σάρξ, flesh; εἶδος, resemblance), the jelly-like material constituting the body substance of the protozoa. **Sarcous**, pertaining to flesh or muscle. **Sarcolemma** (λέμμα, rind), the sheath which encloses a striated muscular fibre.

Scalariform (*scalaria*, stairs; *forma*, shape), ducts found in ferns and club-mosses, with cross markings like the steps of a ladder.

Scaphognathite (σκάφη, a bowl; γνάθος, the jaw), an apparatus in crustacea, for baling the water out of the gill chambers.

Scapula (Lat.), the shoulder blade.

Sclerenchyma (σκληρός, dry, hard; ἔγχυμα, tissue), hard woody fibre; applied also to the calcareous part of a growing coral.

Sclerotic (σκληρός, hard), the hard fibrous coating of the eye.

Secundine (*secundus*, second), the inner coat of the vegetable ovule.

Segmentation (*segmentum*, a piece cut off, from *secare*, to cut), the marking off of anything into parts or segments, as the body or limb of an arthropod; or the substance of the ovum into definite divisions.

Sensation (*sentire*, to feel), feeling; **sensibility**, the faculty of feeling; **sensory**, communicating sensations, as the sensory nerves, and the organs of special sense.

Sepal, one of the leaflets forming the calyx of a flower.

Septum (Lat.) a partition, such as divides the auricles or ventricles of the heart; *septum narium* (*nares*, the nostrils), the partition between the nostrils.

Sertularidæ (*sertum*, a garland), an order of hydrozoa.

Serum (Lat. whey), the fluid part of the blood after coagulation. **Serous membrane**, a membrane whose function is to secrete serum on its surface, as the pericardium, the pleura, arachnoid membrane, etc.

Sessile (*sessilis*, sitting, from *sedere*, to sit), applied to leaves that have no petiole or foot-stalk.

Sinus (Lat. a hollow or recess), a receptacle for venous blood. The **frontal sinus** is an empty cavity in the bone of the skull.

Somatic (σῶμα, the body), pertaining to the body of an animal, as the *somatic cavity*. **Somite**, one of the "segments" into which the body of annulose animals is divided.

Sorus (σωρός, a heap), a cluster of sporangia in ferns.

Spermarium (σπέρμα, seed), the organ in which the sperm corpuscles are produced.

Spheroidal (σφαῖρα, a ball), shaped like a ball.

Spine (*spina*, a thorn or prickle), a modified leaf or branch, bearing a sharp point; applied to the processes of the vertebræ. **Spinal column**, the back bone. **Spinal cord**, the nervous cord which proceeds from the brain along the hollow in the spinal column.

Spiral (*spira*, σπεῖρα, anything coiled), twisted like the thread of a screw, as the spiral striæ in chara, spiral ducts, etc.

Spore (σπορά, seed), a reproductive cell in fungi, *charæ*, ferns, and other cryptogams. Some are produced asexually, some sexually. **Sporangium** (ἀγγεῖον, a vessel), in ferns, the case containing the spores.

Stamen (Lat. a thread), the male organ of a flower, consisting of a filament and an anther, the latter containing pollen grains.

Stellate (*stella*, a star), star-shaped.

Sternum (στέρνον, the breast), in vertebrata, the breast bone; in arthropoda, the inferior pieces of the exoskeleton. **Sternal artery**, the vessel which supplies blood to the inferior parts of the arthropoda. **Sternal sinus**, the cavity which receives the venous blood in the arthropoda previous to its passage into the gills.

Stigma (στίγμα, a puncture), the upper part of the pistil uncovered by epidermis; applied also to the openings of the air-cells in insects.

Stimulus (Lat. a goad, a spur), any cause which may excite a nerve or muscle to action.

Stipule (*stipula*, a straw), little leaflets which are generally found in pairs at the point where an ordinary leaf joins the stem.

Stoma (στόμα, a mouth), air-openings in the epidermis of plants, especially in the leaves.

Striæ (Lat. the flutings of a column), streaks or stripes. **Striated**, streaked.

Style (στῦλος, a pillar, a spike), the part of the pistil between the ovary and stigma.

Supination (*supinus*, face upwards), the placing of the hand with the palm uppermost.

Swimmerets, the limbs of crustaceans which are employed as paddles for swimming.

Symphysis (σύμφυσις, a growing together), the union of two bones which meet each other, as the rami of the lower jaw.

Syncarpous (σύν, together; καρπός, fruit), applied to a pistil formed by the union of several carpels.

System (συνίστημι, to put together), a term comprehending all the parts of the body whose functions are alike, as the *nervous system*, etc. Applied also to the body as a whole, or as distinguished from a particular organ. Hence **systemic**, belonging to the body in general, as the *systemic arteries*, the *systemic circulation*, etc.

Systole (συστίλλω, to compress), the contraction of any of the chambers of the heart.

Tarsus (ταρσός, the flat of the foot), the collection of small bones forming the heel and the part of the foot just under the ankle.

Telson (τέλσον, Homeric form of τέλος, the end), the last joint in the abdomen of the higher crustacea.

Tendons (*tendere*, to stretch), strong bands of fibrous tissue which connect bones with the muscles which move them.

Tentacles (*tentare*, to feel or touch), the finger-like processes surrounding the mouth in cœlenterata, used for feeling and grasping.

Tergum (Lat. the back), the upper part of the ring in the somite of an arthropod. **Tergal**, pertaining to the back.

Ternary (*terni*, three each), consisting of *three*, or some multiple of that number; applied to the number of similar parts in the flower of an endogen.

Thorax (θώραξ, a breast plate), applied in vertebrates to the part of the trunk enclosing the heart and lungs; in arthropoda, to the part between the abdomen and the head.

Tibia (Lat.), the shin-bone, which, with the *fibula*, constitute the *crus*, or shank of the hinder limb.

Transverse section (*transversus*, across; *sectio*, a cutting), a diagram showing the arrangement that would be presented by the parts of an object if it were cut across.

Tubuli (Lat.), little tubes, such as are found in glandular structures.

Tympanum (τύμπανον, a drum), the cavity in the ear behind the **tympanic membrane** or *drum* of the ear.

Ulna (Lat. the elbow), the bone which is felt at the elbow, one of the two which constitute the fore arm.

Umbilicus (Lat. the navel), the point at which the embryo animal draws nourishment into its body.

Umbo (Lat. the boss on a shield), the beak on a bivalve shell.

Unguis (Lat. a claw), the narrow part of a petal by which it is attached to the axis of the flower.

Vacuole (dim. of *vacuum*), an empty space in the sarcode of several of the protozoa.

Vagina (Lat. a sheath), a lateral expansion on the petiole of some leaves, which thus forms a kind of sheath to the stem.

Valves (*valvæ*, folding doors), applied to the shells of *bivalve* molluscs; also to the apparatus within most of the canals of the body, which will allow of the flow of fluid in one direction only.

Vascular (*vas*, a vessel), consisting of vessels. **Vascular tissue**, in plants, contains vessels or ducts. The **vascular system**, in animals, consists of the blood-vessels of the body taken as a whole.

Vena cava (Lat. *hollow vein*), the large vein which discharges the venous blood into the heart; **venous blood** is the dark, impure blood, which requires for its purification to be exposed to the air in the lungs or gills. **Venous sinus**, a cavity in which venous blood is collected in some animals before it is sent to the heart or gills. **Venation**, the arrangement of the "veins" or fibro-vascular bundles in the leaf of a plant.

Ventral (*venter*, the belly), relating to the under part of the body.

Ventricle (*ventriculus*, a little cavity), one of the chambers of the heart; one of the cavities of the brain.

Vertebra (Lat.), a joint of the back, from *vertere*, to turn.

Vertebral column, the back bone. **Vertebral arches**, those that enclose the neural cavity on the upper side, and the thoracic and abdominal cavities on the under. **Vertebrate animal**, one possessed of a back bone.

Verticillate (*verticillus*, the whorl of a spindle), applied to a circlet of leaves all growing from the same point in the stem.

Vesicle (*vesicula*, a little bladder), a little sac. **Vesicular**, consisting of vesicles.

Vestibule (*vestibulum*, an entrance), a chamber in the labyrinth of the ear; one of the canals in the organ of Bojanus.

Vexillum (Lat. a standard), the upper petal of a papilionaceous flower.

Vibrions (*vibrare*, to quiver, little moving filaments developed in organic infusions.

Viscera (Lat. the entrails), applied to all the organs contained in the cavity of the body.

Vitellus (Lat.), the yelk or yolk of an egg.

Vitreous humor (*vitrum*, glass; *humor*, moistness), the substance which fills the ball of the eye, so called from its glassy appearance.

Zoea (ζωή, life), the larval condition of the lobster, and some others of the higher crustaceans.

Zoospores (ζωός, alive; σπορά, seed), spores provided with cilia, and capable of independent movement.

INDEX.

The Numbers refer to the Paragraphs.

ABDOMEN in lobster, 175, 178 ; in vertebrates, 206, 210.
Acacia, 117.
Acephala, 151.
Acetabulum, 223.
Acetic acid, action of on protoplasm, 7 ; on connective tissue, 213.
Achlya, 59.
Acrogen, 65, 77.
Actinia, 122, 142, *et seq.*
Actinozoa, 122.
Adductor muscles, 152, 157.
,, impressions, 158.
Adnate stipules, 117.
Aërial hyphæ, 53.
Afferent nerve, 199 238.
Agamogenesis, 101, 140.
Air passages in plants, 79, 81, 87, 94.
Alæ in bean-flower, 86.
Albumen, 99 ; albuminous seeds, 99.
Alburnum, 112.
Alcohol, 16; action of on cell of chara, 66.
Alternation of generations, 84.
Ambulatory limbs, 176, 178, 180.
Ammonia, a result of the decay of protein, 10.
Amœba, 30, *et seq.*, 121.
Amphibia, 205, 226, 227.
Annual plant, 111.
Annular ducts, 76.
Annulosa, 174.
Annulus, 82.
Anodon, 151, *et seq.*
Antebrachium, 221.
Antennæ ; antennulæ, 181.
Anterior fissure of spinal cord, 236.
Anthers, 86, 97.
Antheridium of peronospore, 57 ; of chara, 63; of fern prothallium, 83, 102.
Antherozoids, 59, 73, 83, 102.
Anura, 205.
Anus, 35, 152, 187.
Aortæ in anodon, 162; in lobster, 189; in frog, 227.
Appendages in chara, 63 ; in fern, 75; in bean, 85, 95, 103.
Appendicular skeleton in lobster, 178; in frog, 217, 219, *et seq.*
Aqueous vapour absorbed by leaves, 81.
,, humour, 243.

Arachnida, 174.
Arachnoid membrane, 234, 239.
Archegonium, 83, 102.
Area pellucida, 244.
Arterial blood, 226, 227.
Arteries in anodon, 162 ; in lobster, 184; in frog, 226.
Arthropoda, 174, 186, 191, 202.
Asexual reproduction, 55, 84, 102.
Ascent of sap, 81.
Aspidium filix mas, 82.
Astacus fluviatilis, 204.
Atlas, 217.
Auricles of heart in anodon, 161; in lobster, 188; in frog, 226.
Axial skeleton, 217.
Axilla, 63, 95.

BACTERIUM, 44, *et seq.*
Baling apparatus of lobster, 181.
Bark, 89, 113.
Barm, 3.
Basipodite, 175.
Batrachia, 205.
Bean, 85, *et seq.*
Bile in anodon, 160 ; in lobster, 187 ; in frog, 224.
Bivalve shells, 157.
Blastema, 11.
Blastoderm, 244.
Bleeding of vines, 107.
Blind spot, 243.
Blood, corpuscles of, 40, 164, 191, 227.
,, circulation of in anodon, 161 ; in lobster, 188, 189, 190 ; in frog, 226, 227.
Blood, inflammatory, 40.
Bojanus, organ of, 153, 163, 188.
Bone, 215.
Brachium, 207, 221.
Bracken, or brake fern, 75.
Bracts, 92, 117.
Brain, 239.
Branches in chara, 63, 72; in bean, 85, 92 ; mode of growth in, 95 ; nature of, 101 ; modified into spines or tendrils, 117.
Branchial chamber in anodon, 154.
,, cavity in lobster, 176, 181.
,, heart of tadpole, 246.
Branchiostegite, 176.

Branchlets in chara, 63.
Brandy, manufactured from woody fibre, 7.
Bronchi, 225.
Brownian movement, 46.
Bud developed into a branch, 95; analogous to fern spore, 102; terminal, 89; axillary, 95; scales of, 117.
Budding, 12, 126.
Bundles, fibro-vascular, 87, 88, 115.
„ of muscular fibre, 230.
Bulbus arteriosus in lobster, 189; in frog, 227.
Byssus of young anodon, 171.

CACTUS, 104.
Calyx, 86.
Cambium, 88, 112.
Camphor, movements of in water, 46.
Capillaries, 163, 226.
Carapace, 176.
Carbonate of lime in stem of chara, 66; in shell of anodon, 156; in bone, 215.
Carbon in plants, 104.
Carbonic acid absorbed by protococcus and all green plants, 26, 27, 79, 80, 103; resolved by chlorophyll into carbon and oxygen, 26, 27, 63, 79, 80, 104, 105; evolved by torula, 15, 16; by penicillium, 54; by all fungi, 60; by green plants, 27, 80; by animals, 36, 132, 227; results from oxidation of protein, 10.
Carina, 86.
Carotid arteries, 227.
Carpel, 86, 98.
Carpus, 221.
Cartilage, 214; intervertebral, 217.
„ bone, 216
Casein, 9.
Caulinary stipule, 117.
Caustic potash, action of on protoplasm, 7, 22, 52.
Cell, 6, et seq.
Cell wall, 6, 7.
Cellulose, composition of, 7.
Centra of vertebral column, 209, 217.
Central canal of spinal cord, 236.
Cephalic flexure, 176.
Cephalo-thorax, 176.
Cerebellum, 239.
Cerebral ganglia, 168, 197; hemispheres, 240; nerves, 241.
Cerebro-spinal axis, 233.
Cervical suture, 176.
Chara, 63, et seq.
Chelæ, 180.
Chitin, 32, 141.
Chlorophyll present in protococcus and all green plants, 20, 26, 27, 67, 76, 94, 103; absent from torula and all fungi, 20, 60; breaks up carbonic acid into carbon and oxygen, 26, 27, 63, 79, 80, 104, 105.
Chondrin, 214.
Chorda dorsalis, 244.
Choroid coat, 243.
Cilia, 21, 28; in spores of fungi, 57; in hydra, 130; in actinia, 143; in anodon, 159, 160, 166; in antherozoids, 59, 73, 83, 102; in sperm corpuscles, 141, 170, 171, 244.
Ciliary process, 243.
Circulation in hydra, 132; in actinia, 144; in anodon, 161, et seq.; in lobster, 188, et seq.; in frog, 226, et seq.
Clavicle, 216, 220.
Cleavage of yelk, 244.
Cloaca in anodon, 152, in frog, 224.
Cnidæ, 123.
Coagulation in amœba, 39.
Coccyx in frog, 217.
Cochlea, 243.
Cœlenterata, 122, et seq.
Columella auris, 243.
Commissural cords, 153, 168.
Condyles of skull in frog, 217.
Confervæ, 62.
Conjugation, 58.
Conidia, 52.
Conidiophores, 53.
Connate stipules, 117.
Connective tissue, 213.
Continuous gemmation, 140.
Contractility in amœba, 38, 41; in protococcus, 28, 41; in chara, 68; in hydra, 126, 129, 132, 134, et seq.; in actinia, 143, 145, et seq.; in all muscular tissue, 229.
Coracoid bones, 220.
Coralligena, 150.
Coral reefs, 150.
Corky layer in bark, 113.
Cornea of lobster, 200; of frog, 243
Corolla, 86.
Corpora bigemina, 240; C. striata, 240.
Cortex, 88.
Cortical part of stem in chara, 66.
Cotyledons, 99, 100, 119.
Craspedum, 143.
Cray-fish, 204.
Crus in frog, 223.
Crust of lobster, 175.
Crustacea, 174.
Ctenophora, 123.
Cucumber, tendrils of, 117.
Cyclosis in chara, 68.

DEHISCENCE of sporangium, 82.
Dermis, 228.
Diaphragm, 210.
Diastole, 31.

Diatomaceæ, 31.
Dicotyledones, 120.
Differentiation, 112, 114, 133, 136, 167, 213, 228, 230, 244.
Digits, 207, 221, 223.
Dionæa muscipula, 117.
Disc, 135, 143.
Dischidia Rafflesiana, 117.
Discontinuous gemmation, 140.
Dorsal side of flower, 86.
,, aorta, 227.
Ducts in plants, 76, 79, 81, 87, 88, 90, 94; in animals, 187, 203, 224, 228, 229.
Ductless glands, 40.
Dura mater, 233.
Duramen, 112.

EAR of anodon, 169; of lobster, 201; of frog, 243.
Ectoderm in hydra, 123, 127, *et seq.*; in actinia, 143.
Efferent nerves, 199, 238.
Electricity, action of on amœba, 38; on cyclosis in chara, 68.
Embryo in chara, 73; in fern, 83; in bean, 99; in all plants, 120.
Embryo of lobster, 203; of cray fish, 204; of frog, 245.
Embryo sac, 98.
Embryonic spot, 244.
Emulsion, 224.
Encephalon of frog, 239.
Encystation of amœba, 32.
Endoderm in hydra, 123, 127, *et seq.*; in actinia, 143.
Endogens, 114, 115, 116, 118, 119, 120.
Endogenous cell multiplication, 203, 244.
Endolymph, 243.
Endophlœum, 113.
Endophragmal partition in lobster, 183.
Endopodite, 175, 180, 181.
Endorhizal germination, 120.
Endoskeleton in frog, 213.
Endosmosis in plants, 81, 107.
Endosperm in fern spore, 82; in seeds, 98, 99.
Ependyma of brain, 239.
Epiblast, 244.
Epidermis in fern, 76; in bean, 87, 88, 90; in shell of anodon, 156, 157; in frog, 209, 228.
Epiblema of root, 90.
Epibranchial chambers in anodon, 153.
Epiphlœum of bark, 113.
Epipodite in lobster, 175, 180, 181.
Epithelium in anodon, 159, 160, 166, 170; in lobster, 186; in frog, 209, 228.
Exalbuminous seeds, 99.

Excretory apparatus in hydra, 133; in anodon, 160, 165, 166; in lobster, 187, 193; in frog, 227.
Exhalant aperture in anodon, 170.
Exhalation of carbonic acid by plants, 27, 80; of moisture, 81.
Exogenous plants, 85, 114; stems, 89, 90.
Exopodite, 175, 180, 181.
Exorhizal germination, 120.
Exoskeleton in anodon, 152, 156; in lobster, 175.
Exosperm in fern spore, 82.
Extine of a pollen grain, 97.
Eye of lobster, 181, 200; of frog, 243.

FACETS in cornea of lobster, 181, 200.
Fascia of muscles, 230.
Fascicles, 116, 120.
Fat, 8, 9.
Fecundation (or fertilization) of oögonium, 57, 59; of nucule, or spore fruit, 73; of archegonium, 83; of ovule, 99; of ovum, 141, 171, 203, 244.
Feelers of lobster, 181.
Female organs of flower, 101; of fern prothallium, 102.
Femur, 207, 223.
Fenestra ovalis, 243.
Fermentation, 16.
Fern, 75, *et seq.*, 102.
Fertilization (see *Fecundation*).
Fibre, woody, 7, 76, 87, 88, 90, 94, 101, 103, 113.
Fibre, muscular, 129, 135, 143, 167, 194, 229, 230.
Fibre, nerve, 169, 198, 236.
Fibrillæ of muscle, 194, 230; of nerve, 169.
Fibrin, 9, 165.
Fibro-vascular bundles, 87, 88, 114, 115.
Fibula, 223.
Filament of stamen, 86.
Fission in protococcus, 23; in penicillium, 55; in chara, 70; in actinia, 143.
Fissures in spinal cord, 236.
Flax, fibres of, 113.
Flower, 86, 96, *et seq.*; not borne by ferns, 102.
Foot of anodon, 152.
Foot-jaws of lobster, 176.
Foramen of ovule, 98, 99; intervertebral, 237.
Foraminifera, 121.
Fore-brain, 240.
Formative material, 11, 14.
Forces at work in the propulsion of sap, 106.

Fovilla, 102.
Fresh-water polype, 121, *et seq.*
  ,,   mussel, 151, *et seq.*
Frog, 205, *et seq.*
Frond of fern, 75, 78, *et seq.*, 93.
Fungi, 43, 52, *et seq.*; reproduction in, 61.
Funiculus of ovule, 86.
Furze, spines of, 117.

GALL bladder in frog, 224.
Gamboge, movements of in water, 46.
Gamogenesis, 55, 101, 126.
Ganglia of nerve system in anodon, 153, 168, 169; in lobster, 183, 197; in frog, 236, 237, 242.
Ganglia cerebral, 153, 168; pedal, 168; parieto-splanchnic, 168.
Ganglionic corpuscles, 169, 198, 236.
Gastric juice, 131, 224.
Gelatine, 213.
Gemmation in torula, 12; in hydra, 126; in coralligena, 140.
Gemmation continuous, 140; discontinuous, 140.
Gemmules of actinia, 149.
Germ cell (or corpuscle), 59, 61, 141.
Germinal cell (or vesicle), 99; spot, 141; area, 244.
Germination, endorhizal, 120; exorhizal, 120.
Gills of anodon, 151, 154; of lobster, 176, 181, 190, 191, 192; of tadpole, 246.
Glands, 181, 187, 193, 224, 231, 240, 243; ductless, 40.
Glandular sac of organ of Bojanus in anodon, 163.
Glenoidal cavity, 220.
Glochidia, 172.
Glottis, 225.
Gluten, 9.
Glycerine in fermented sugar, 16.
Glycogen, 41, 42.
Gooseberry prickles, 117.
Green bark, 114.
  ,,   colour of plants due to chlorophyll, 94, 103.
  ,,   glands in lobster, 181, 193.
  ,,   mould (a fungus), 52.
Grey matter of spinal cord, 236; of medulla oblongata, cerebellum, and cerebral hemispheres, 240.
Gristle, 214.
Growing point in chara, 65, 70; in exogens, 98, 109; in endogens, 114.
Guard cell in stoma, 79.
Gum, 7, 107.

HÆMAL aspect of body, 153.
Hallux of vertebrates, 223.

Hard fern, 82.
Harderian gland, 243.
Hay, infusion of, 44.
Heart in anodon, 161, *et seq.*; in lobster, 184, 188, *et seq.*; in frog, 209, 210, 211, 226, 227, 229, 231.
Hermaphrodite actinozoa, 150.
Hilum of bean seed, 100 (see fig. 55, A.)
Hind brain of vertebrates, 240.
Hinge in anodon, 152 (see *Ligament*).
Hollow in bean stem, how formed, 87.
Hollow muscles, 229, 231.
Homarus vulgaris, 174, *et seq.*
Homology in chara, 64.
House-leek, 104.
Humerus in vertebrates, 220, 221.
Humus, 104.
Hydra, 122, *et seq.*: *fusca*, 126; *viridis*, 126.
Hydrozoa, 122, 124.
Hymenium of mushroom, 60.
Hyoid bone in frog, 224.
Hyphæ, 52, *et seq.*; aërial, 53.
Hypœsophageal ganglia in lobster, 197.
Hypoblast, 244.

ILIUM in frog, 222.
Impregnation, 59 ; of ovule, 99.
Incus in vertebrates, 243.
Indusium in ferns, 81.
Inflammatory blood, 40.
Infundibulum, 240.
Infusoria in hay infusion, 50.
Innominate bone in frog, 222.
Inorganic materials in plants, 104.
Insecta (insects), 117, 174.
Insertion of a muscle, 167.
Inspissation of sap in leaves, 81, 106.
Intercellular spaces in leaves, 79, 81, 87, 94, 103.
Internodes in chara, 63; their mode of growth, 70, 71 ; in fern, 77, 78 ; in bean, 85.
Intervertebral cartilages, 217 ; foramina, 233, 237.
Intestines in anodon, 159, 160; in lobster, 185, 186; in frog, 212, 224, 229; in tadpole, 246.
Intine of pollen grain, 97.
Intussusception, 11, 29.
Involuntary muscles, 231.
Iodine colours protoplasm, 7, 44, 52 its action on cellulose with sulphuric acid, 22.
Iris of the eye, 243.
Irritability of hydra, 126, 134 ; of actinia, 143, 145.
Irritation of roots of nerves, 238.
Ischium in frog, 222.
Iter a tertio ad quartum ventriculum 240.

INDEX.

JELLY fishes, 122.
Joints of lobster, 194, *et seq.*

LABIUM and labrum in lobster, 185.
Labyrinth of ear in frog, 243.
Lacteals, 224.
Lacunæ in lobster, 190; in bone tissue, 215.
Lamellæ, 151; of mushroom, 60.
Lamellibranchiata, 151.
Lamina of frond, 78; of leaf, 117; of gills in anodon, 154.
Lamina terminalis in brain, 240.
Laminæ dorsales in vertebrate embryo, 244.
Laminations in shell of anodon, 156.
Larval hydra, 141; actinia, 149; anodon, 172; lobster, 203; frog, 246.
Lateral ventricles of brain, 240.
Leaf, structure of, 94; mode of growth of in chara, 70; in fern, 78; in bean, 93; modified into a tendril, spine, or pitcher, 117.
Leaflets in chara, 63.
Leaf-stalk, 92.
Leptothrix, 47.
Leucocythæmia, 40.
Liber of bark, 113.
Ligament in anodon, 157; in frog, 217, 220.
Light not necessary to existence of torula, 14; aids chlorophyll in decomposing carbonic acid, 26, 63, 79, 80, 103, 104.
Lignine, 112, 213 (see *Woody fibre*).
Limb of petal, 86; of lobster, 175, *et seq.*; of frog, 207.
Lime, carbonate of, in shell of anodon, 156; in exoskeleton of lobster, 195; phosphate and carbonate of, in bone, 215.
Liver in anodon, 160; lobster, 187; frog, 224.
Lobe of mantle in anodon, 152.
Lobster, 174, *et seq.*
Locomotion in hydra, 135; actinia, 146; anodon, 167; lobster, 194, *et seq.*
Lymph corpuscles, 40.
Lymphatic vessels in man, 40; in frog, 224.
Lymphatic system, 224.
  ,, hearts, 224.
Magenta stains protoplasm, 7, 22, 33, 52.
Male apparatus of flower, 97, 101; of fern, 102.
Male element in peronospore, 58; in achlya, 58.
Malleus of ear in vertebrates, 243.
Mammalia, 210.
Mandible in lobster, 181; in frog, 218, 224.

Mandibular palp, 181; ramus, 218.
Mantle of anodon, 152; cavity, 152 (see *Pallium*).
Manubrium, 73.
Matrix in cartilage, 214.
Maxillipedes, 176, 180.
Medulla in bean, 87.
  ,, oblongata, 239. 240.
Medullary part of stem in chara, 66.
  ,, rays, 68, 111.
Melon, tendrils of, 117.
Membrana nictitans, 206, 243.
Membrane bones, 215.
Mesenteries of actinia, 125, 143.
Mesentery in frog, 211.
Mesoblast, 245.
Mesophlœum, 113.
Metacarpus of frog, 221.
Metamorphosis, 11, 29, 77.
  ,, of anodon, 171, 172; lobster, 203; frog, 246.
Metastoma in lobster, 185.
Metatarsus in frog, 223.
Micropyle, 98, 99.
Mid-brain, 240.
Middle bark, 113.
Midrib, 92; modified into tendril, 117.
Mineral salts in protein, 9; in bean, 104.
Mobile protococcus, 20, 21, 28.
Moisture absorbed and exhaled by leaf, 81.
Mollusca, 151, 164.
Monocotyledones, 120.
Morphology, 4.
Motor nerves, 239.
Mulberry mass in fertilised ovum, 244.
Multiplication by budding, 12, 82, 101, 140, 148; by fission, 23, 34, 47, 55, 148; sexual, 55, 57, 58, 83, 73, 99, 101, 126, 141, 149, 170, 202, 244.
Muscular fibre in hydra, 123, 129; in actinia, 123, 143; in anodon, 167; in lobster, 194, 196; in frog, 229, *et seq.*
Mushroom, 60.
Mussel (fresh water), 151, *et seq.*
Mycelium, 52, 53.
Myriapoda, 174.

NACRE in shell of anodon, 156.
Nematocysts, 123.
Nepenthes, 117.
Nerves, absent in hydra, 136; in actinia, 145; afferent and efferent, 199, 238.
Nerve cells and fibres, 169, 198, 236.
Nerve trunks, 237, 238; roots, 237, 238.
Nervous system in anodon, 168, 169; lobster, 197, *et seq.*; frog, 233-241.
Neural arches, 209, 244; N. aspect of body, 158; N. cavity of a vertebrate, 209.

Neurilemma, 199, 237.
Nictitating membrane in frog, 206, 243.
Nitrogen a constituent of protein, 9.
Nodes in chara, 63, 64; mode of growth in, 70, 71.
Notochord, 244.
Nucleus in amœba, 31; in colourless corpuscle of blood, 40; in cell of chara, 67, 68, 69, 74; of fern, 76; of cœlenterata, 123; of anodon, 167; of lobster, 198; of frog, 213, 214, 215, 228.
Nucleus of vegetable ovule, 98, 99.
Nucule in chara, 63, 73.

ŒSOPHAGUS, 153, 197, 224.
Œsophageal collar, 153, 197.
Olfactory lobes in frog, 240.
Oögonium of peronospore, 57.
Oöspore, 57.
Operculum of gills in tadpole, 246.
Optic lobes, 240; thalami, 240; nerves, 241, 243.
Organ of Bojanus in anodon, 153, 163; functions of, 165.
Organic matter in soil not essential to life of plant, 104.
Origin of a muscle, 167.
Os innominatum, 222; pubis, 222; quadratum, 218.
Oscillatorie, 45.
Osmunda Regalis (Royal Osmund), 82.
Otoconia in ear of lobster, 201.
Otoliths, 169, 243.
Ovary in anodon, 170; in lobster, 203.
Ova in hydra, 141; anodon, 171; lobster, 203; frog, 244.
Ovules, 86; development of, 98, 99; fecundation of, 99.
Oxidation of tissues, 80, 132, 192, 227.
Oxygen in a free state not essential to torula, 15.
Oyster, development of, 173.

PALLIAL cavity, 152; impression, 158; muscles, 167.
Pallium of anodon, 152.
Palpi labial, 153; mandibular, 181.
Pancreas in frog, 224.
Pancreatic juice, 224.
Papilla of embryonic anodon, 171.
Parallel venation of endogens, 118.
Parenchyma in fern, 76; in bean, 87, 88.
Parieto-splanchnic ganglia 168.
Pasteur's fluid, 3, 12.
Patella (knee cap), 223.
Pea, tendrils of, 117.
Pectoral arch in frog, 220.
Pedal ganglia in anodon, 168.
Peduncle of fern frond, 78, 93; of lobster's eye, 200.

Pelvic arch in frog, 222.
Penicillium, 52, *et seq.*
Perennial exogen, 111.
Pericardium in anodon, 161, 164; lobster, 184, 188; frog, 211.
Perilymph in ear of frog, 243.
Periosteum, 233.
Peristaltic action of heart in anodon, 162.
Peritoneum in frog, 212.
Perivisceral cavity, 125, 143.
Peronospora infestans, 56; its propagation asexually, 56; sexually, 57.
Petals of flower, 86, 96.
Petiole of leaf, 92, 117.
Phalanges of frog, 221.
Phanerogams, 99, 110.
Phosphate of lime in bones, 215.
Phyllodium (phyllode), 117.
Physalia (Portuguese man-of-war), 128
Physiology, 4.
Pia mater, 234, 235.
Pileorhiza of root, 103.
Pileus of mushroom, 60.
Pincers of *Chelæ*, 180.
Pineal gland of brain, 240.
Pinnules of fern frond, 78.
Pistil of flower, 86, 101.
Pitchers and pitcher plants, 117.
Pith, 87, 88.
Pituitary body (in brain), 240.
Placenta in vegetable ovule, 96, 98.
Plasma, 3, 18, 30.
Pleura of vertebrates, 211, 212.
Pleuron of lobster, 175.
Plexus of nerves, 235.
Plumule of vegetable embryo, 99.
Pneumogastric nerves, 241.
Pollen grains, 86; development of, 97, 99.
Polyhedral cells, 76.
Polypes, 122, 150.
Pons Varolii, 240.
Portuguese man-of-war, 128.
Portal vein, 247.
Posterior fissure of spinal cord, 236; in medulla oblongata, transformed into a simple concavity, 239.
Post-œsophageal ganglia, 197.
Potassium, salts of, in protoplasm of torula, 9; in bean, 104.
Potato blight, cause of, 56.
Prickles, modified epidermis, 117.
Primine of vegetable ovule, 98, 99.
Primordial utricle, 67.
Primitive groove in ovum of frog, 243.
Procœlous vertebræ of frog, 217.
Proëmbryo in chara, 73.
Pronation of *manus*, 221.
Propagation of plants, sexual, 99, 101, 102; asexual, 101, 102.

Protein a constituent of protoplasm, 8; composition of, 9; waste of, 37.
Prothallium of fern, 83, 102.
Protococcus, 18, *et seq.*, 79, 103; contains chlorophyll, 26, 37, 103; decomposes carbonic acid, 26, 27, 79; type of green plants, 62, 79, 103; nivalis, 24.
Protoplasm, 6, 8, 20, 31; composition of, 8; stained by iodine or magenta, 7, 22, 33, 44, 52.
Protopodite in lobster, 175.
Protovertebræ in embryo frog, 244
Protozoön (*pl.* protozoa), 30, 121.
Protractor pedis in anodon, 167.
Pseudopodium, 31, 40.
Pulmo-cutaneous artery in frog, 227.
Pulmonary circulation, 226.
Pupil of eye, 243.
Putrefaction induced by bacteria, 48.
Pylorus of stomach, 185.

QUADRATE bone, 218.
Quaternary and quinary arrangement of parts in flowers, 119, 120.
Quiescent protococcus, 20.

RADICLE of embryo plant, 100, 120.
Radius, one of the bones of the forearm, 221.
Rain water, contents of, 19.
Ramus of mandible, 218.
Recti muscles of eye, 243.
Rectum intestine in anodon, 159.
Red corpuscles of blood, 40, 227.
Red snow (protococcus nivalis), 24.
Renal excretions in anodon, 165; in lobster, 193; in frog, 224, 227.
Reproduction in torula, 12; protococcus, 23; amœba, 34; bacterium, 47; penicillium, 55; peronospora, 56, 57; achlya, 59; fungi, 61; chara, 73; fern, 82, 83, 84; bean, 99-102; hydra, 139-141; actinia, 148, 149; coralligena, 150; anodon, 170-172; oyster, 173; lobster, 202, 203; cray-fish, 204; frog, 244-246.
Respiration in plants, 79, *et seq.*; in hydra, 132, 133; in actinia, 146; in anodon, 166; in lobster, 191, 192; in frog, 225.
Reticulated venation in exogens, 118.
Retina of eye, 243.
Retractor muscles in anodon, 153, 167.
Rhizoids in chara, 63.
Rhizome of fern, 75.
Ribs of vertebrates, 209.
Rigor mortis, 230.
Rods and cones in eye of lobster, 200.
Root, structure of, 90; mode of growth in, 91.

Root and root stock of fern, 75.
Roots of spinal nerve, 237
Rootlets, 63, 90.
Rotifers in hay infusion, 50.
Rostrum of lobster, 176.
Royal Osmund fern, 82.
Rythmical action of heart, 162; of *bulbus arteriosus*, 189, 227; of *sinus venosus*, 227.

SAC (or cell wall), 6; composition of, 7.
Saccharine solution fermented by torulæ, 3, 17.
Sacral bone (or vertebra), 217, 235.
Salivary glands absent in frog, 224.
Sap, ascent of, 81, 106-108; elaborated in leaves, 81, 109.
Sapwood (alburnum), 112.
Sarcode, constitutes body of protozoön, 121.
Sarcolemma of muscle, 230.
Sarcous elements in muscles of anodon, 167.
Sarracenia, 117.
Scalariform vessels in fern, 76, 77
Scaphognathite of lobster, 181.
Scapula in pectoral arch of frog, 220.
Sclerenchyma in fern, 76, 77.
Sclerotic coat of eye, 243.
Sea-anemone, 122, 142, *et seq.*
Secundine of vegetable ovule, 98, 99.
Seed of plant, 99, 100.
Segmentation (cleavage) of yelk, 244.
Semicircular canals in ear, 243.
Sensibility in hydra, 126, 134, 136, 137; in actinia, 143, 145.
Sensory nerves, 199, 238.
 ,, organs, in anodon, 169; in lobster, 199-201; in frog, 243.
Sepal of flower, 86, 96.
Septum dividing chambers of heart, 216.
Sertularidæ, 122.
Serum of blood, 225.
Sessile leaf, 117.
Sexual propagation, 57; in fern, 84; in bean, 99, 100; in hydra, 141; in actinia, 149 (see *Reproduction*).
Shell of anodon, 152, 156, 158
Shield fern, 82.
Sight, organs of in lobster, 199, 200; in frog, 243.
Sinus venosus (or venous sinus) in anodon, 163; in lobster, 190; in frog, 227.
Skeleton of anodon, 152, 156-158; of lobster, 175-181; of frog, 217-223.
Skin of frog, 208; aids in respiration, 225, 227.
Skull of frog, 217.
Smell, organs of in frog, 243.

Smooth muscular fibre in anodon, 167; in vertebrates, 186; in frog, 229, 231.
Somatic cavity in cœlenterata, 123-125.
Somites in lobster, 175, *et seq.*
Sorus in fern, 82.
Spat of oyster, 173.
Sperm corpuscles, or cells, in achlya, 59; chara, 73; hydra, 141; anodon, 170, 171; lobster, 202; frog, 244.
Spermarium of lobster, 202.
Spermatozoa, 141.
Spinal cord, 235; nerves, 235.
Spines, modified leaves or branches, 117.
Spines in vertebral column of frog, 217.
Spirælæ (bacteria), 45.
Spiral ducts in fern, 76; in bean, 87, 88; in endogens, 115.
Spiral striation in chara, 66, 87.
„ turning in stem of bean, 87.
Sponges (protozoa), 121.
Spontaneous generation, 51.
Sporangium of fern, 82.
Spore fruit in chara, 63, 72, 73.
Spores of bacterium, 49, *et seq.*; of penicillium, 55; of fern, 82, 84, 102.
Squamosal bone of mammals, 218.
Stamens of flowers, 86, 97.
Stapes in ear of frog, 243.
Starch, 7, 9, 22, 76, 107.
Stellate form of nerve corpuscles, 236.
„ order of cells in leaf, 94.
Stem of chara, 63, *et seq.*; fern, 75, *et seq.*; bean, 85, 87, *et seq.*
Sternal artery in lobster, 184, 189; sinus, 190.
Sternum of lobster, 175; of frog, 209, 220.
Stigma of flower, 86, 96, 97, 90.
Still protococcus, 20.
Stimulus, action of, on hydra, 134; on actinia, 143; on muscle, 229.
Stipules, 92, 117.
Stomach of anodon, 159, 160; lobster, 185, 186; frog, 224.
Stomata, 79, 81, 87, 94.
Striated (or striped) muscular fibre, 167, 186, 194, 229, 230, 231.
Striation, spiral in chara, 66, 87.
Style of flower, 86, 90, 99.
Sub-aërial hyphæ, 53.
Suberous layer in bark, 113.
Succinic acid, 16.
Sulphur a constituent in protoplasm of torula, 9.
Sulphuric acid, action of, on cellulose with iodine, 22.
Sunlight, action of, on carbonic acid, with chlorophyll, 26, 27, 62, 79, 80, 104.

Supination of hand, 221.
Sweet pea, stigma of, 97.
Sympathetic nerve system, 209, 242.
Symphysis of mandible, 224.
Syncarpous pistil, 86.
Systemic circulation in frog, 226.
Systole, 31, 191, 227.

TADPOLE, 246.
Tarsus of foot, 223.
Teeth of lobster, 185; of frog, 224.
Telson of lobster, 175.
Tendrils of vine, 117.
Tentacles of cœlenterata, 123; of hydra, 127, 128, 132, 135, 140, 141; of actinia, 143, 144, 145, 149.
Tergum of lobster, 175.
Terminal bud in chara, 70; in fern, 77, 78; in bean, 89, 91, 93.
Ternary arrangement of parts in endogenous flowers, 119, 120.
Thorax in lobster, 184; in vertebrates, 209, 210.
Thorn, a modified leaf, or branch, 117.
Thread cells, 124, 128.
Tibia in limb of frog, 223.
Tongue in frog, 224.
Torula, 3, *et seq.*
Trachea (windpipe), 225.
Tradescantia, cyclosis in hairs of, 68.
Trigeminal nerves, 241.
Tubules in liver of anodon, 160; of lobster, 187.
Tympanic membrane of ear, 243.

ULNA in limb of frog, 221.
Umbilicus (or navel), 204.
Umbo of valve in anodon, 156.
Unguis of petal, 86.
Urea excreted by hydra, 132.

VACUOLE, 6; in amœba, 31; in cells of chara, 67; of hydra, 130.
Vagina of leaf, 117.
Valves of anodon, 152, 156, 158, 171.
Valves of heart, 169, 188; of nasal passages in frog, 206, 225.
Vascular air passages, 103.
Vegetative frond in fern, 82.
Veins in leaf, 92, 117, 118, 120; modified into spines or tendrils, 117.
Veins in frog, 226, 227.
Vena cava (or venous sinus) in anodon, 163; lobster, 190, 191; frog, 227.
Venation in exogens and endogens, 118, 120.
Venous blood, 163, 191, 226, 227.
Venus's fly-trap, 117.
Ventral side of flower, 86.
Ventricle of heart in anodon, 159; lobster, 188; frog, 226, 227.

INDEX.

Ventricle of brain in frog, 240.
Vertebral arches, 209.
Vertebrate animals, 205, *et seq.*
Verticillate leaves, 63.
Vesicle, germinal, 90.
Vessels, dotted and spiral, 87, 88, 94.
Vestibule in organ of Bojanus, 163.
Vexillum in bean flower, 86.
Vibrions, 45.
Vines, bleeding of, 107; tendrils of, 117.
Viscera, 123, 175, 209.
Visceral cavity of vertebrates, 209.
Vitelline membrane, 172.
Vitellus (or yelk), 244.
Vitreous humour of eye, 243.
Vomerine plates, 224.

Wall of cell, 6, 128; of body cavity, 124, 125, 143, 150, 211.

Water a result of the oxidation of protein, 10.
Water flea, 126.
Wax covers aërial hyphæ, 53.
Web of frog, 221, 227.
White matter of spinal cord, 236; of cerebellum and cerebral hemispheres, 240.
Whorls in chara, 63; fern, 75; bean flower, 86, 90.
Winged petiole, 117.
Woody fibre, 7, 76, 87, 88, 111, 114, 115.
Wolffian bodies, 227.

Yeast, 3, 10.
Yelk, 173, 204, 244.

Zoea of lobster, 203.

www.ingramcontent.com/pod-product-compliance
Lightning Source LLC
Chambersburg PA
CBHW021733220426
43662CB00008B/826